W9-CVZ-238

FISH MIGRATION

FISH MIGRATION

BRIAN A. McKEOWN

Department of Biological Sciences,
Simon Fraser University,
Burnaby,
British Columbia,
Canada

CROOM HELM
London & Sydney

TIMBER PRESS
Portland, Oregon

1984

Croom Helm Ltd, Provident House, Burrell Row,
Beckenham, Kent BR3 1AT
Croom Helm Australia Pty Ltd, First Floor,
139 King Street, Sydney, NSW 2001, Australia

British Library Cataloguing in publication Data

McKeown, Brian A.
Fish migration.
1. Fisheries – Migration
I. Title
597′.052′5 QL639

ISBN 0-7099-1761-9

First published in the USA in 1984 by
Timber Press,
PO Box 1632,
Beaverton, OR 97075,
USA

ISBN 0-917304-99-3

Printed and bound in Great Britain

CONTENTS

For MERRILY

and

JANICE, SHARON, JAMES AND DAVID

PREFACE

The subject of fish migration has not been reviewed in book form since the work of Harden-Jones in 1968. This work was a major contribution but did not cover the area of physiology as it applies to the migration of fish. Since Harden-Jones' book other recent works have been published such as those by Baker (1978, 1982) and Gauthreaux (1980). These books deal with the migration of animals in general and the latter also covers physiological aspects. However, the more detailed aspects related specifically to fish are often lacking. Woodhead (1975) has written an extensive review on endocrinological aspects of fish migration and Leggett (1977) has reviewed the ecology of fish migrations.

It thus appeared to me that it would be helpful to review most aspects of fish migration specifically and bring this topic more up to date. My research is mostly concerned with the physiology of Pacific salmon migration, particularly the control of energy reserves as well as osmotic and ionic regulation. Chapters concerned with fish physiology may thus be emphasized more than others as they reflect my personal background. In no way does this bias reflect my perceived importance of other aspects of fish migration. For the sake of completeness, I have tried my best to cover the topics of orientation, bioenergetics, migratory patterns, behavior, ecology and evolution. I hope this coverage makes my description of the phenomenon of fish migration more complete and to the satisfaction of most readers.

Much of the research information available concerning physiological, as well as other aspects of fish migration are gathered from laboratory studies. It might, however, seem more meaningful if various relevant parameters could be investigated from fish while actually migrating. Yet, to elucidate the underlying physiological mechanisms, controlled laboratory studies are required. Thus, much of this book will appear to be focused on laboratory conducted experiments into the elucidation of specific physiological mechanisms. However, answers to pertinent questions on how fish can migrate are often only obtainable by these types of studies and as such are included in this book.

Many aspects of animal migration are not unique to fish although some features such as olfactory orientation may be more important to fish or better understood in particular fish species than in other animal groups. There is a wealth of information on how animals migrate obtainable from studies on a wide variety of taxonomic groups. This book focuses on the migration of fish and the information available from other animal groups is largely omitted, due to limitations on space.

This book would not have been possible without the advice, encouragement, data and reviews by many people. They are: L. Dill, R. Fargher, G. Geen, T. Hardwick, J. Neilson, T. Quinn, R. Sweeting, G. Wagner and M. Walker.

I would also like to thank Ms J. Rosenberg and the Instructional Resource Center at Simon Fraser University for the preparation of most figures. I extend my special thanks to Ms W. Yee for typing the manuscript and helpful suggestions concerning format. Finally, my gratitude to my wife Merrily and children is hard to express. They were most understanding of the time spent and distractions from family activities. Because of them and for them, the whole undertaking was worthwhile.

1 INTRODUCTION

Migration – a Problem of Definition

For countless generations man has marvelled at the annual returns of migrating fish. The knowledge that many of these animals have often travelled great distances, encountered many obstacles or predators and managed to seek out and find specific sites has inspired great admiration. Over the ages man has developed many explanations, theories and superstitions regarding these miraculous fish migrations. Although man has expected the annual return of specific fish populations and indeed in many cases depended upon such movements for his own survival, the answers to many questions still elude the lay observer as well as the scientific researcher.

When individuals are asked to describe a typical fish migration they usually envisage a species that displays a number of oustanding characteristics. Thus, a migrant fish is visually obvious due to the sheer mass of numbers. Indeed many fish may exhibit schooling behavior when they migrate and such concentrations become obvious to the casual observer and are exploited by the commercial fisherman. Spawning migrations of salmon into fresh water or large schools of herring in the oceans are such examples. Many fish migrations become evident because of their precise timing. Some upstream migrations of adults or downstream migrations of juveniles to feeding areas occur at very predictable times of the year, especially if they are photoperiodically induced. If such seasonal movements are synchronized by temperature fluctuations, a certain degree of predictability is still inherent, and the passing of the seasons, therefore, makes such fish movements conspicuous.

The migration of certain fish species may impress the observer because of the resilience of such fish. These fish meet the demands of sometimes long and arduous journeys both energetically and physiologically. Many fish move across great expanses of open ocean or move, from far out to sea, long distances up difficult inland waterways. These migrations often cover distances of thousands of kilometers and thus require refined adaptational abilities. Not only the distance but also the speed of migration can be equally impressive. For example, some bluefin tuna migrate from Florida across the Atlantic ocean to the coast of Norway, a straight line distance of over 10,000 kilometers. This transatlantic crossing takes from two to four months during which time individuals can average speeds of over 5 km/hr. Other species, particularly those that migrate upstream, display movement patterns requiring great endurance. Not only do they swim against the current but they must also negotiate a number of obstacles such as waterfalls. To observe the determination and ability of a large migrating salmon attempting to overcome a waterfall many times higher than the length of the fish is an impressive sight.

Some fish migrations are notable due to the precision of their return. Certain species migrate long distances but return very specifically to a particular site from whence they migrated. The ability of salmon to home to their natal area is remarkable when one thinks of the number of years some salmon have been away from the spawning site and the various distant environments encountered in the interim.

Migrating fish have evolved many general as well as specific morphological, behavioral and physiological adaptations. Fish that travel long distances, especially in the open ocean, must be capable of orienting to specific directions given the appropriate environmental cues. This ability of a fish to find its way to a specific goal, at a precise time and the capability to recognize that goal as the correct destination is characteristic of certain migrations. Many fish move between fresh water and sea water and thus require the ability to osmoregulate or adjust to changes in salinity. This adaptation, plus many others such as acquisition, storage and mobilization of energy reserves are all important for many fish species in order to accomplish a migration. In many cases such adaptations are in fact essential for migration.

The preceding are just a few examples of characteristics of migrating fish. However, no single species exhibits all the above characteristics. Some species may be capable of great speeds during migration, others may swim slowly but travel far, yet others may traverse the freshwater/sea water boundary, whereas some move only within the freshwater or marine environment, etc. Thus, any given fish species may be notable because of some outstanding single characteristic of its movement pattern. Certain features of migrations in general may be absent from those of specific species or may be present as only a minor constituent. Some fish species display so few aspects of general migration that one might not even recognize or define that species as a migrant. Indeed, since there are gradations in any one or all possible characteristics of a migrant fish, an all-encompassing definition becomes difficult if not almost impossible. On the other hand the inclusion of only specific characteristics to describe migratory behavior would also appear unsatisfactory because of biases in making the definition. Faced with this dilemma it would thus seem difficult to put forth any particular definition for the term 'migration'. However, a working definition will be attempted here in order that a logical discussion can ensue to describe actual movement patterns of fish. It is hoped that this definition will lead the reader to an appreciation of the scope and diversity of such patterns. It is also anticipated that a discussion of this term will lead to an understanding of how and why fish move from place to place, and although information is still incomplete it is hoped that the reader will appreciate the prodigious undertakings and adaptations displayed by migrant fish.

The general meaning of the word migration is to move from one place to another. In the past, migration of animals has had the connotation of a coming and going with the seasons or an implied periodicity. Landsborough Thompson (1942) used the term migration to describe rather specific types of movement patterns. He noted that many animals move from one local area to another with the changing seasons. They cover relatively short distances and he refers to these as simply local or seasonal movements and not migratory behavior. Other animals may cover long distances, but if they are only in one direction, Landsborough Thompson would class them as dispersals. He defines a true migration as a seasonal movement that implies some element of a return to some initial starting point. Harden-Jones (1968) discusses migration of fish according to the definition of Heape (1931). Migration is thus 'a class of movement which impels migrants to return to the region from which they have migrated'. This is similar to earlier definitions in that it implies a return movement and in addition some factor(s) driving the migration. 'Impels' is not meant to be taken anthropomorphically but refers to parameters impinging on the biological well-being of the species or individual. These types of migrations are classed as *alimental* in that they are for the purpose of food procurement, *climatic* for purposes of reaching a region of better climate or *gametic* for purposes of reproduction.

As previously outlined, the different types of animal movements overlap and often exhibit distinctions only on purely arbitrary grounds. Some fish such as herring move vertically on a daily basis in order to feed. At night they are found on the surface feeding and during the day they are in deeper, colder water. This movement may have come about to avoid predators during the day or possibly to conserve energy by being in colder waters when not feeding. These are return movements based on time of day but not based on the time of year or season and therefore by some authors this would not be considered a migration. Other fish species may move from feeding area to feeding area due to food availability that changes with the seasons. However, some such fish may not return to original areas but instead continually move off to new locales and thus might not qualify as a migration. However, either of these examples might be considered a migration depending on the definition. Baker (1978) adopts a more general and all-encompassing definition. He suggests that arbitrary restrictions have previously been imposed on the term migration and defines migration as 'the act of moving from one spatial unit to another'. There are no restrictions as to return movements or confinements within time. Spatial unit implies that the distances covered may be small or great and might be traversed by groups or by individuals. Indeed it might even refer to movements within an individual such as cell migrations. Such a definition might also even apply to inanimate objects.

In past decades the phenomenon of migration was treated and defined as a movement pattern with special restrictive conditions, but later a more general and all-encompassing phenomenon was envisaged. More recently, the term migration has been further modified, with only slight but pertinent restrictions which I personally find most meaningful and applicable to many animal movement patterns. This definition is enunciated by Dingle (1980) and reads as follows: 'Migration is specialized behavior especially evolved for the displacement of the individual in space.' This definition is similar to that of Baker (1978) with one salient difference. The 'specialized behavior' implies that there are also specialized physiological mechanisms underlying the appropriate behavior patterns. Dingle's definition implies evolutionary changes leading to specific migratory patterns or mechanisms. Thus, as pointed out by Dingle, accidental or unintentional movements cannot be acted on by natural selection and therefore should not be included in a definition of migration. With this definition in mind I have tried to discuss various aspects of fish movement. Patterns of migration (Chapter 2) are discussed for two main reasons. One is to attempt a description of movement patterns in order to develop an appreciation of the various general patterns of how fish move about in fresh water, the oceans or between the two environments. Secondly, the patterns discussed often show overlapping as well as unique characteristics. The movement of a fish from place to place is obviously an energetically costly business. The ways and means by which a fish meets these energy demands are discussed in Chapter 4. The migration of fish requires specialized functions to have been developed depending on the species and the environmental parameters encountered on the journey. Physiological adaptations (Chapter 5) are thus required for such processes as osmoregulation, metabolite mobilization, sensory adaptation, reproduction, and specialized behavior patterns. Migration patterns have evolved, are maintained or continue to change due to the development of specialized orientation mechanisms. These mechanisms (Chapter 3) give fish the ability to find their way from place to place and recognize their final goal once it is attained.

Diverse and impressive patterns of fish migration exist today and some information is available to help us understand orientation, bioenergetic and physiological mechanisms. However, changes in or the evolution of such patterns and mechanisms from previously non-migrating species or species with different migratory behaviors also need to be considered. The reasons why fish migrate and the evolution of such behaviors will thus form the concluding chapter for this book.

Migration Direction

Migrations exhibit a number of different characteristics that vary between different species as well as between individuals within the same species. This and the following sections discuss some of these characteristics in order to identify similarities and differences among groups of fishes as well as between fish and other animal groups.

Many animals migrate with the seasons and move along a more or less north-south geographical axis. Fish also migrate along this axis but display considerably more variation than other groups of animals and migrations exist which follow virtually all geographical directions. Terrestrial migrants move in basically a two dimensional or horizontal plane. Such migrants may also change their position with respect to altitude and might thus be considered as moving in three dimensions, but at any given altitude the animal is on one particular plane that extends in two dimensions. Birds in flight and fish in the aquatic environment have the option of changing their movements to another plane and migrate continuously in a three dimensional world. This gives such animals the added advantage of choosing different environments for migration. The vertical axis within a water column, for example, may have very different currents to assist fish movements. Celestial cues may differ with depth, creating vast differences in orientation possibilities. Temperature, salinity, light intensity, light polarization, light quality, predation, food availability and other parameters all change with depth of water and thus add multivariate characteristics to the space in which a fish can choose to migrate.

When a specific population of fish migrates from one place to another it is tempting to derive a compass direction between these two places and subsequently describe this particular population as orienting exactly along this specific bearing. However, when fish are tested for a direction preference during migration, individuals differ and the whole population migrates in a mean direction with variations around this average. Baker (1969) uses the term *direction ratio* to describe this variation in the scatter of movement directions about a mean direction. Basically this ratio expresses the percentage of individuals within a population that move in the different compass quadrants relative to the mean direction of movement. A few examples will be considered to point out differences that appear to exist between some fish with respect to movement directions (Figure 1.1). The European eel, *Anguilla anguilla*, has been suggested as having a spawning ground south of Bermuda and west of the Bahamas, the so-called Sargasso Sea. The young larvae of these fish move north-east and enter freshwater streams along the coast of Europe where they feed with adults. They start their migration at a single point but spread out over a direction of approximately 45° about the mean direction. The American eel, *A. rostrata*, is probably a separate species to the European eel and might also breed in the Sargasso Sea but slightly more southerly (see Chapter 2). The young of this population move towards the east coast of North America and enter fresh water from the Gulf of Mexico north to Newfoundland, which is a direction of approximately 120° about the mean direction. The pink salmon, *Oncorhynchus gorbuscha*, feeds in the north Pacific ocean but displays spawning migrations to Japan, USSR, Alaska, British Columbia, Washington, Oregon, and California, which represents a scatter of directions of over 200°. Some species such as the herring, *Clupea harengus*, and the plaice, *Pleuronectes platessa*, that feed and spawn in the North Sea, show migration directions towards breeding grounds that might be scattered around a total of 360°.

When considering the migration of an individual fish over great distances the impression might be that the fish is very accurate in homing towards a precise location. Indeed this might be the case for individuals but when different races or populations within a given species are considered, not all individuals arrive

Figure 1.1: Direction ratios for fish populations of different species. (a) European eels, *Anguilla anguilla*, are thought to migrate as young from the spawning grounds somewhere in the Sargasso Sea to various river systems in Europe; a direction ratio of approximately 45°. (b) American eels, *Anguilla rostrata*, may also migrate from the same area or nearby to the European eels. They, however, migrate to river systems along the eastern coast of North America; a direction ratio of approximately 120°. (c) Pink salmon, *Oncorhynchus gorbuscha*, feed in the North Pacific Ocean before migrating to rivers in Asia or North America to spawn; a direction ratio of approxmately 200°. (d) Many oceanic migrants such as some populations of herring, *Clupea harengus*, and plaice, *Pleuronectes platessa*, migrate to breeding areas that may be scattered in almost all directions (360°).

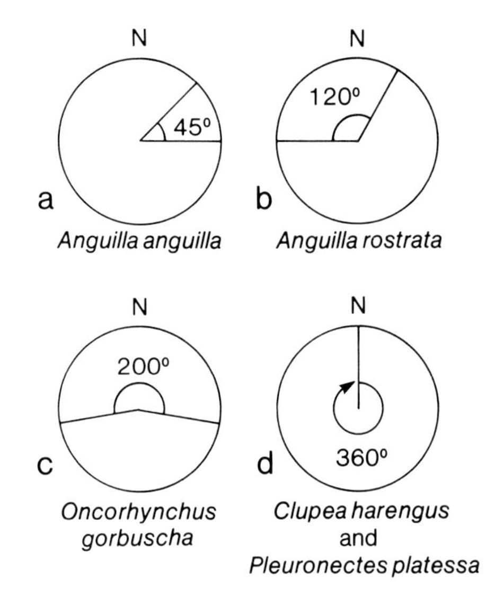

at the same place and the direction ratio becomes a means to assess the degree of orientation of a group. Thus, some populations may show very little scatter in direction of movement whereas others may be less precise. As the populations become larger and larger, so also does degree of scatter. The information on direction ratio is useful when contemplating issues such as orientation, homing, speed of migration, environmental influences and evolution of migration patterns. The determination of direction ratio thus allows for a quantitative description of the amount of bias shown by a group of migrating fish towards the mean direction.

Batschelet (1965) has used other terms which allow for a quantitative comparison between movement directions of migrating fish. This latter direction parameter is referred to as the mean vector and is the mean direction of a series of directions. The mean vector has direction as well as length. The length of the mean vector is a measure of the bias towards the mean direction. If a population of fish moves away from some point in direction evenly distributed around 360° there would be no mean direction and the mean vector would have a length of zero. If all fish moved off in the same direction the direction of the mean vector would be established and it would have a length of 10. These descriptions are particularly useful for analyses of orientation behavior (see Chapter 3 – magnetic field orientation).

Periodicity

As already discussed, migration is a difficult concept to define due to the gradation of most parameters that characterize it. When fish move from place to place each group of individuals usually varies in the direction taken in these movements. Indeed, even the definition of place is difficult. A 'different' place ecologically can be defined at all levels of gradation and the boundary line between places therefore becomes imprecise. Baker (1978) has suggested that a different place might be one outside sensory contact. However, confusion again arises as to what constitutes sensory contact. Time-compensated sun-compass orientation may give a fish sensory contact anywhere on the globe. A salmon migrating upstream orients to olfactory cues emanating from the natal area. It would be difficult to imagine that upstream movements of salmon are not a display of migratory behavior. Periodicity of fish migrations also exhibit an array of movements with respect to time. Some Pacific salmon may thus remain at sea for several years (up to eight years for Chinook salmon, *Oncorhynchus tshawytscha*) before returning to the spawning grounds. Larvae of the sea lamprey, *Petromyzon marinus,* may spend several years filter-feeding in the mud on stream bottoms before metamorphosing and migrating to sea. Eels also spend several years (up to 20) feeding in fresh water before spawning migrations to the sea begin. Other fish species migrate annually, monthly or daily – such as the vertical migrations of herring. Spawning migrations of the Pacific salmon species may be observed annually, but individuals undertake upstream spawning migrations only once in their life cycle and even often at different ages from other salmon of the same species. The periodicity thus varies between different species as well as within species. Other differences occur between species since many fish such as Atlantic salmon, *Salmo salar*, can spawn more than once such that the periodicity may vary for individual spawnings.

Most migratory fish exhibit a single migratory periodicity that may be confined to daily movements for some species or individuals but might also be extended over many years for other fish. Some fish species display migration patterns based on a number of environmental parameters and consequently have different sub-categories of periodicities. The California grunion, *Leuresthes tenuis*, is a good example of a migratory species that moves according to a number of changing variables. The grunion has an annual spawning migration towards the shallow inshore areas of California throughout the spring and summer. These fish spawn high up on sandy beaches well out of the reach of potential marine predators of their newly developing offspring. In order to reach the highest levels on the beach, these fish migrate during the highest tides. These higher tides occur at different phases of the moon so that migrations up onto the beach occur on a monthly or semi-monthly basis. The spawning adult fish have the opportunity to migrate during the appropriate phase of the moon but would be at a disadvantage if they exposed themselves during the day as the risk of predation would be high. Therefore, not only do these fish require the appropriate season of the year and the right lunar phase or tide level, but also they migrate only at night, thus imposing a third variable.

Most fish migrations occur at regular intervals whether they be daily, monthly, seasonally or for longer periods such as yearly, biennially or longer. These migrations may appear to be due to initiating factors that are mainly biological in nature. Thus, fish migrate to their spawning areas when gonadal recrudescence is at an appropriate stage or move to feeding areas due to an increase in food availability in the new area or a decrease in food availability in their present area. These examples of biological parameters leading to regular migratory behavior are the result of ultimate factors, whereas the proximate factors imposing the movement are probably abiotic in nature (for example, change in temperature, photoperiod, salinity, lunar cylces, precipitation, currents, etc.). Some fish migrations are irregular in that they occur at variable intervals. These might be considered as 'irruptions' by some authors, although large numbers of individuals migrating at regular intervals when normally only a few migrate might also be referred to as irruptions. Such irregular migrations might also depend on abiotic factors such as with the movement of lung fish into the mud to aestivate when their freshwater pond dries up. On the other hand, some migrations might only occur when the fish reaches a certain condition such as the premigratory fattening stage. Other irregular events may also occur based on what seem to be biotic factors such as food availability or reduction in predator numbers.

Distance, Speed and Duration

Although the distances covered by different migrant fish vary considerably, it is a component of migration that affects both the pattern of movement as well as the various required mechanisms of orientation, bioenergetics, physiology and behavior. Exceedingly long migrations include those of salmon, tuna and eels, while short migrations include the daily vertical migrations or inshore/offshore feeding movements of some fish. Other short distance migrations seem rather uninteresting to some observers, but from the point of view of the particular species involved, the obstacles overcome in traversing the relatively short distance may still be impressive. The small inshore Gobiid, *Bathygobius soporator*, has amazing specializations to migrate short distances. It has the ability to relocate to tide pools above the surface in the intertidal zone when the tide is out. When relocating to such a pool it jumps out of the water. If it lands high and dry it will obviously suffer physical damage or be exposed to a high risk of predation. This is a relatively short distance to travel, yet this fish requires a considerable physical ability to jump the required distance and to be capable of orienting in the required direction. Successful orientation is thought to depend on prior knowledge of the topography learned when the tide is high. When these vastly different migration distances are put into perspective each fish is seen to be capable of remarkable migratory accomplishments irrespective of the distance traversed.

The speed, duration and thus distance of any fish migration depends on the type of environment through which the fish must pass as well as the environmental requirements sought in the final destination. Many fish migrations are for the purpose of spawning. Sometimes individual fish must migrate great distances in order to locate appropriate sites for reproduction. From start to finish, fish may traverse environments of varying suitability. For example, fish that migrate long distances require large amounts of energy to sustain the necessary swimming. Often these fish are also going through a stage of gonadal development during migration and thus have an added requirement for an energy supply. The environment that the migrant fish passes through on its way to the spawning site might be able to provide a food source so that the procurement of energy is possible along the way. Other fish migrants encounter harsh environments where the energy expenditure in finding, capturing, and assimilating food is greater than the energy gained. These fish must therefore acquire and store energy reserves before the onset of the migration and then draw upon these resources while migrating. The distances covered by the two species might be similar but the time taken to complete the distance might differ if the fish have to gather food along the way.

Other migrations are undertaken for the purpose of locating better feeding areas and such examples may help to illustrate causes of variations in fish migration distances. Some fish move into an unknown environment and are somehow able to test the suitability of that environment to provide an adequate food supply, which if adequate may temporarily end the migration until the food becomes scarce. Baker (1978) describes this migration distance as facultative.

Other fish species migrating to a feeding area may adopt a different strategy. If the migration is of a facultative nature, the fish must spend time testing and assessing the suitability of each new environment. The chance of encountering a suitable environment may be low and the distance required to reach that new environment may be great. Thus, some fish may migrate a distance which may be termed obligatory. That is, if a particular fish moves in a given direction over a specific distance, it is likely to encounter a suitable environment. These fish must, however, have prior information of the requirements necessary to reach this suitable area or at least have inherited the appropriate genetic information to undertake a migration that is obligatory in nature. The possibility also exists whereby a fish may perform a migration over a facultative distance and once having discovered a suitable environment might return to it a second time by migrating over an obligatory distance.

Degree of Return

Most concepts of fish migrations involve some degree of return movement, although as defined earlier migration need not necessarily involve a return component. When the patterns of various fish migrations are described, it is evident that there is a gradation from very precise return movements to a one-way migration (Figure 1.2). Any fish that exhibit a high degree of return frequent many areas more than once during the course of the migration. Some return migrations, however, have only one area that is frequented more than once. In most return migrations, the breeding site is the area returned to most precisely. Many Pacific salmon, for example, are known to return precisely to the same spawning area where they themselves were hatched. The feeding areas in the ocean, however, are much more diverse and individuals in succeeding generations may feed in quite different locations. Repeat spawners that home

Figure 1.2: The degree of return for different fish migrations. (a) Fish that migrate from A to E and never revisit any one place. This may be uncommon during the entire life of an individual fish, but may occur in some salmon as an example. Most salmon return to natal areas to spawn but some 'wander' and spawn in new areas, thus having no common area during their migration. (b) A to-and-fro pattern of return. For example, many trout in fresh water migrate upstream to spawn and thence retrace their route back to feeding areas. (c) A circular type of return migration that is exemplified by many marine fish migrants that follow ocean currents flowing in a circular fashion. (d) A combination of a to-and-fro and circular pattern of return. Many diadromous species exhibit this type of migration. The freshwater phase of the migration is the to-and-fro movement along some river course whereas the circular pattern occurs while in the marine environment.

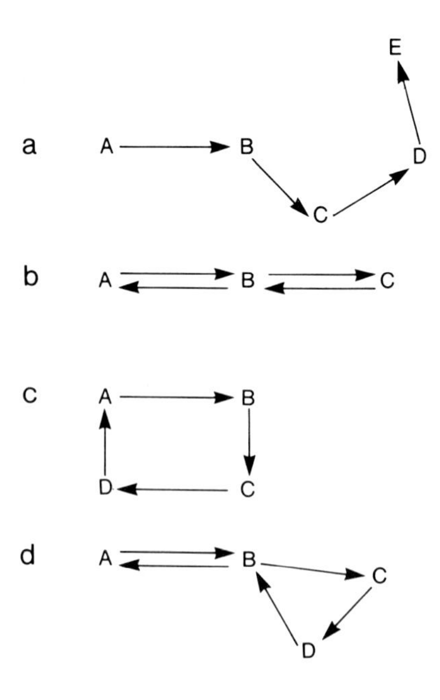

to natal areas, such as many trout species, probably have varying degrees of precision for their return migrations. This situation necessitates succeeding generations to compete for similar areas and some individuals will thus be displaced varying distances thereby changing the accuracy of their return.

Catadromous fish reside in fresh water during their growth to adult phase, then migrate and spawn in the ocean. *Oceanodromous* fish spend their life cycle in the marine environment and also spawn in the ocean. Many such fish spawn in open water and although the spawning areas are broadly defined, it is unlikely that an exact position is relocated at each spawning. The European eel spawns in the south-west portion of the North Atlantic Ocean known as the Sargasso Sea and the spawning locations probably vary considerably from year to year. The adult stage of these eels is spent in various rivers of Western Europe and the young are not likely to ascend the same rivers that their parents did. This species therefore, has a much less precise degree of return for both the feeding and spawning areas than many other species.

The degree of return is largely a function of the general patterns of migration exhibited by fish. A return migration may be accomplished by the fish reversing the direction in which it travelled on the outward-bound part of its migration. This to-and-fro pattern of fish migration may be important for orientation mechanisms or may come about because of restricted areas of movement as in stream migrations. Since this type of migration involves back-tracking along a route travelled before, the degree of return is high and in addition the return is through a number of different areas. Some fish migrations, especially those in the open ocean, follow patterns of the currents and only bring the fish back to one common area. This circular type of migration may thus display a high degree of return to this one area and the loop may be accomplished in either the horizontal or vertical plane. Water currents may be circular in nature, thus forming a gyral. All the fish need do is follow the current to complete the return migration. Other fish may follow a current in the outward journey and then change the plane of movement by ascending or descending in the water column to an area where there is a current in the opposite direction and, by following this current, manage to return to the point of migration initiation. Many fish also display combinations of these types of patterns, an example being migrating salmon. They migrate downstream to the sea and there complete an ocean migration circuit whereupon they re-enter the stream and retrace their route to the natal area. Such migrants must be capable of remembering the correct stream after an interval of several years.

Methods for Studying Fish Migrations

Harden-Jones (1968) has given a comprehensive review of this topic, but as in other contemporary sciences, new techniques have permitted major advances in the field. Techniques used to study fish movements have provided a great deal of insight into descriptions of the migratory patterns of fish. When in these studies the numbers of fish moving from place to place is correlated with changing environmental conditions, additional information becomes available for a better understanding of the strategies used by fish for migration. Thus, the descriptive patterns of migration can often be used to gain ecological insights. A better understanding of orientation mechanisms has often been obtained by descriptive studies of fish movements. If fish are observed to be moving in a specific direction by whatever technique is available, and in a particular environment where the various physical and chemical parameters are known or measurable, actual usable cues for orientation may be elucidated. By following adult migratory Pacific salmon as they approach their river system, it has been shown that movement patterns are correlated with tidal currents and river/ocean current interactions (Madison *et al.*, 1972; Stasko *et al.*, 1973, 1976; Groot *et al.*, 1975). Other experiments by Groot (1965) with sockeye salmon, *Oncorhynchus nerka*, smolts indicated that these fish could orient in specific directions on cloudy days, at dusk or in artificially covered conditions. These experiments comparing orientation ability as correlated to different environmental factors led Groot to postulate a non-celestial compass orientation.

Observations of migrating fish in their natural environment have led to new insights, but controlled laboratory experiments have also been useful in extending our knowledge of particular aspects of migration. Although field observations give useful behavioral data, additional ethological information has been obtained under controlled experimental conditions. Thus, for example, valuable information has been gathered from field observations on young stream-resident juvenile coho salmon, *Oncorhynchus kisutch*, with respect to territoriality and consequential displacements of subordinates (Dill *et al.*, 1981). The territory size of these fish was inversely related to the density of benthic food in the territory (Figure 1.3). Thus, when benthic food becomes scarce, territory size of individuals increases. Under these conditions the observation area contained fewer salmon. The remainder presumably migrated to less desirable locations or grouped together in schools elsewhere due to the assumed intraspecific competition. The above field observations led the investigators to question what behavioral mechanisms might be used by these young coho salmon to adjust their territory size to the amount of food available. A fish might first assess food availability by the degree of fullness of its gut or some other measure of hunger. The effect of hunger on aggression may then influence the establishment of an appropriate territory size. The most relevant aggressive behavior to measure might be the reactive or attack distance towards a territorial intruder. To test this question concerning territoriality Dill *et al.* (1981) had to rely on a controlled experimental environment. Results indicated that the distance from which a resident coho attacked an approaching model intruder increased asymptotically with hunger (Figure 1.4). This example shows that the combination of field and laboratory studies can lead to a better understanding of the behavioral mechanisms underlying migration.

Figure 1.3: The relationship between juvenile coho salmon territory size (square metres) and the abundance of potential benthic prey items in the territory.

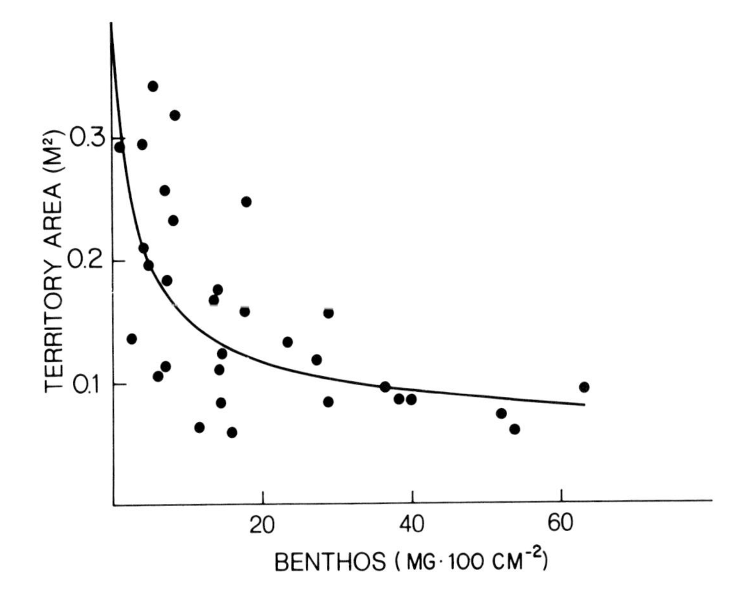

Source: from Dill *et al.*, 1981

Figure 1.4: The relationship between the hunger level of juvenile resident coho salmon and the distance from which they attacked the model intruder.

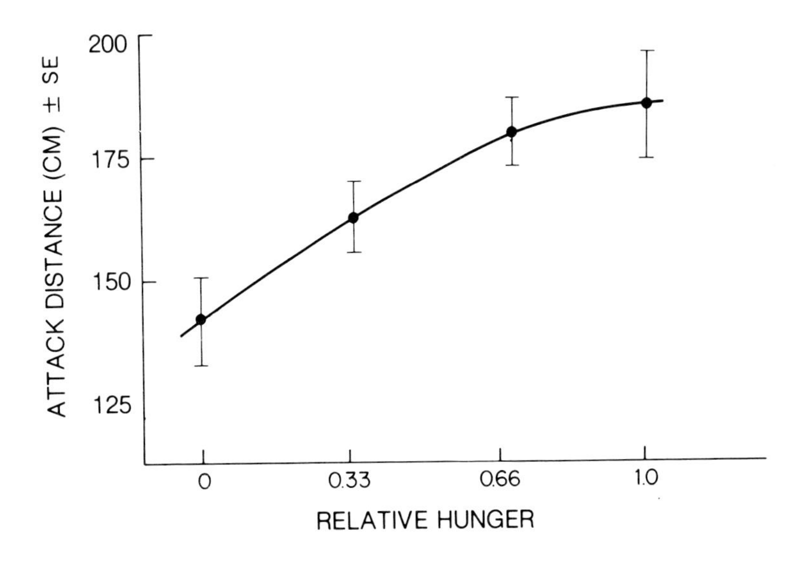

Source: from Dill *et al.*, 1981

The combination of field studies with laboratory investigations using advanced techniques has led to a better understanding of the physiological bases of fish migration. As an example, field observations by commercial fishermen have suggested that adult migratory sockeye salmon move in and out of fresh water several times as they approach the entrance of the Fraser River, British Columbia, Canada. Although the boundary between fresh water and sea water in the Fraser River estuary is difficult to determine, the fresh water is sufficiently more turbid and observations are thus possible to indicate movements between different salinity environments. The migratory fish may be testing Fraser River water for sensory cues but may just as likely be adjusting physiologically to the reduced salinity of fresh water prior to the upstream migration. As will be discussed later (Chapter 5), the anterior pituitary hormone prolactin appears to be responsible for the freshwater survival of several species of fish. To test whether or not adult migratory sockeye require prolactin for survival in fresh water, McKeown and van Overbeeke (1972) collected blood and pituitary samples from migrating sockeye that were far from the Fraser River mouth (360 km), near the entrance to the Fraser River, well within the river (320 km) and on the spawning ground (600 km). When the fish entered fresh water the prolactin content of the blood and pituitary gland was reduced (Figure 1.5). During the subsequent migration to the spawning grounds, blood prolactin levels increased considerably and pituitary levels were restored to previous sea water levels. In order to conduct such studies new sensitive ligand assays had to be developed to measure specifically the low hormone levels which are found in blood. Although tissue samples were collected from fish during the course of their natural migration from the sea into fresh water, a number of difficulties of interpretation arise. Some of the changes in hormone levels could be correlated with the time of entrance of the sockeye into fresh water, but other environmental parameters might also have been changing at the same time. Thus, when the fish encountered fresh water they might also have experienced temperature changes, current-mediated variations in swimming activity, feeding or metabolite modifications, orientation alterations, etc. Due to so many variables, controlled laboratory experiments were required. McKeown and Brewer (1978) transferred sea water acclimated coho salmon directly to fresh water under controlled conditions and compared blood and pituitary prolactin levels to those of other fish transferred from sea water back to sea water as controls for the effects of handling and the transfer *per se* (Figure 1.6). However, even these controlled experiments are difficult to interpret. A drop in pituitary prolactin levels upon transfer to fresh water may be an indication of enhanced secretion into the blood. It may also be an indication that synthetic activity of the prolactin cells is reduced. Electron microscopical examination of prolactin cells from salmon transferred to fresh water suggests that prolactin is released into the blood following exposure to fresh water (Leatherland and McKeown, 1974). The uptake of tritiated leucine, which is one of the major amino acids in prolactin, into prolactin cells is increased if fish (McKeown and Brewer, 1978)

Figure 1.5: Changes in pituitary and serum prolactin. Concentrations in adult sockeye salmon migrating from the ocean to their spawning grounds at Chilko Lake, Canada. Numbers below group numbers indicate distance (miles) along migration route between collection sites and the mouth (arrow) of the Fraser River (–, sites at sea; +, upstream sites). Shaded columns represent lengths of fish (upper histograms) and serum protein concentrations (lower histograms).

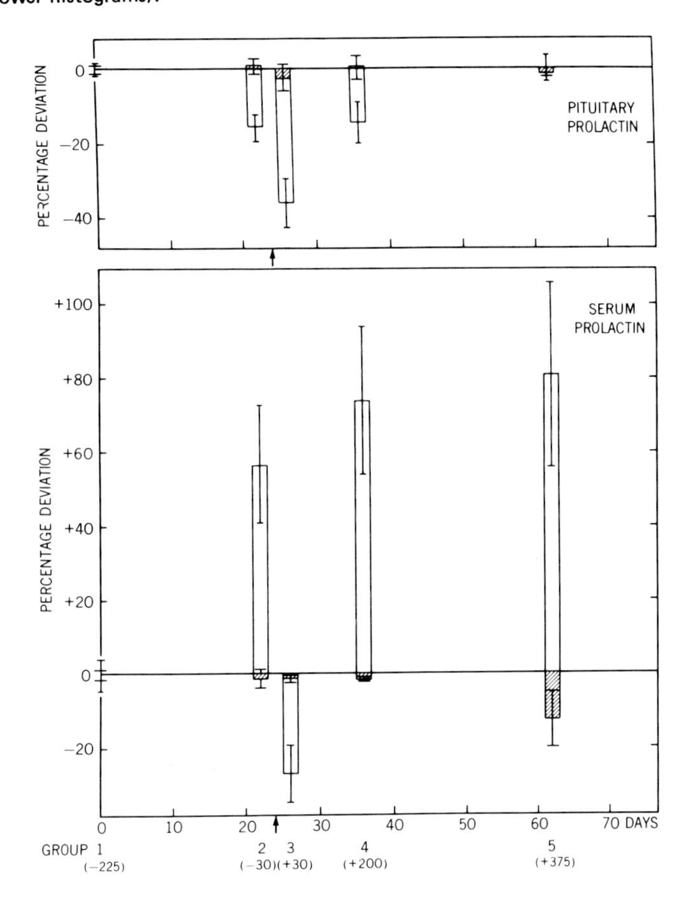

Source: from McKeown and van Overbeeke, 1972

(Figure 1.7) or prolactin cells (McKeown, unpublished) are exposed to water of lower osmotic concentration. These studies and an autoradiographic study of prolactin cells of salmon exposed to fresh water and injected with tritiated leucine indicate an enhanced synthetic rate of these cells in fresh water (McKeown and Hazlett, 1975). The pituitary thus appears to secrete prolactin into the blood upon exposure to fresh water, but the blood levels of prolactin do not reflect an increase at this time and instead show a decrease (Figure 1.6). This is likely to be due to enhanced utilization by the target tissues or breakdown

Figure 1.6: Plasma and pituitary prolactin levels in juvenile coho salmon following an abrupt transfer from sea water to fresh water. Open circles represent values from fish transferred back into sea water whereas closed circles represent values from fish transferred into fresh water.

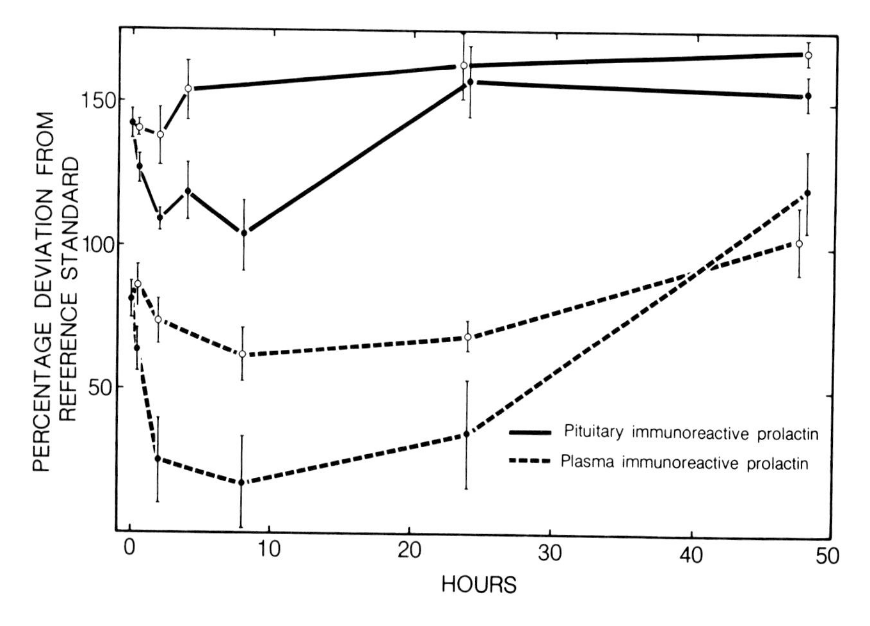

Source: from McKeown and Brewer, 1978

by other tissues. Support for this view comes from a study by Brewer and McKeown (1980) where they showed a shorter half-life of injected ^{125}I-labelled ovine prolactin in salmon transferred to fresh water as compared to sea water fish (Figure 1.8). The combination of these field and laboratory techniques have thus been useful for determining a particular physiological phenomenon during a fish migration.

Marking and Tagging Techniques

Marking and tagging of fish followed by recapture has been a convenient technique employed for almost a century. These identification marks are invaluable for describing patterns of fish migrations and can provide information on rates and directions of fish movements. Mark and recapture methods are also useful for giving estimates of population sizes.

To be useful a mark or tag must stay with the fish until the end of the study period. It is often desirable for the mark or tag to be identifiable enough so that individual fish may be recognized as they migrate. Some tags and marks are either cumbersome or physically harm the fish. These are undesirable as they directly or indirectly harm the fish or make the fish more conspicuous to predators.

Figure 1.7: Tritiated leucine incorporation into protein in pituitary prolactin cells of juvenile coho salmon subjected to a transfer from sea water to fresh water or back into sea water.

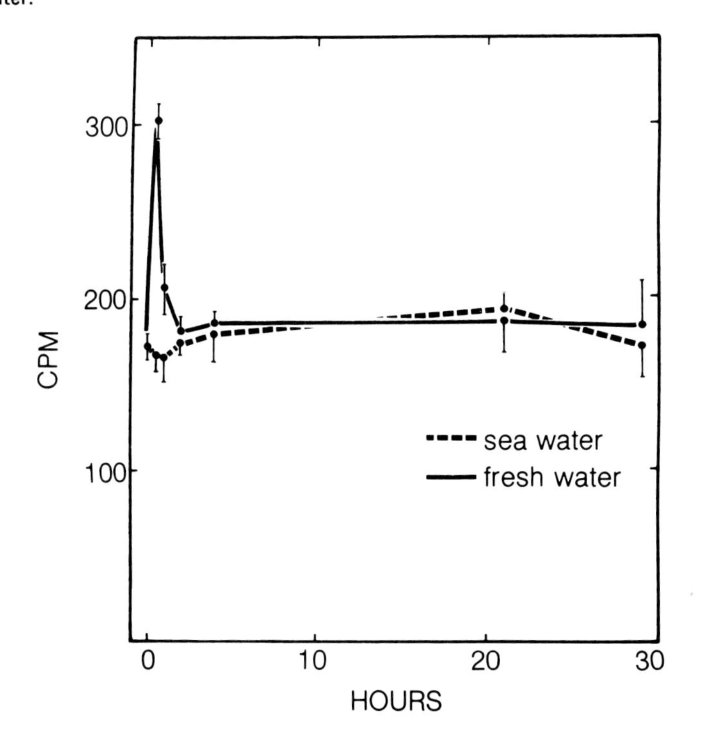

Source: from McKeown and Brewer, 1978

There has been a wide variety of marks applied directly to fish for identification purposes. Both small and large fish have been marked by fin clipping. By using different combinations, a number of different identification marks can be obtained by this method. This is a quick and easy procedure but allows for only a limited number of individuals to be recognized. As a single mark for a large group of individuals it is however very useful. The removal of fins as a marking technique may be complicated by the fact that some fish may regenerate lost fins or may lose fins due to natural causes. To extend the number of combinations, even the outer extension of the maxillary bones of the jaw have been removed. These clipping procedures also have problems because of consequent decreases in fish survival.

Other methods have been developed whereby fish can be tattooed (Hickling, 1945) or branded with metal tools cooled with liquid nitrogen (Figure 1.9) (Mighell, 1969) or heated (McNicol and Noakes, 1979). Lasers have also been used for branding. The marks can be any variety of symbols so that combinations

Figure 1.8: Plasma ^{125}I-prolactin counts in juvenile coho salmon subjected to a gradual change from sea water to fresh water or transferred from sea water back to sea water after receiving an injection of ^{125}I-prolactin.

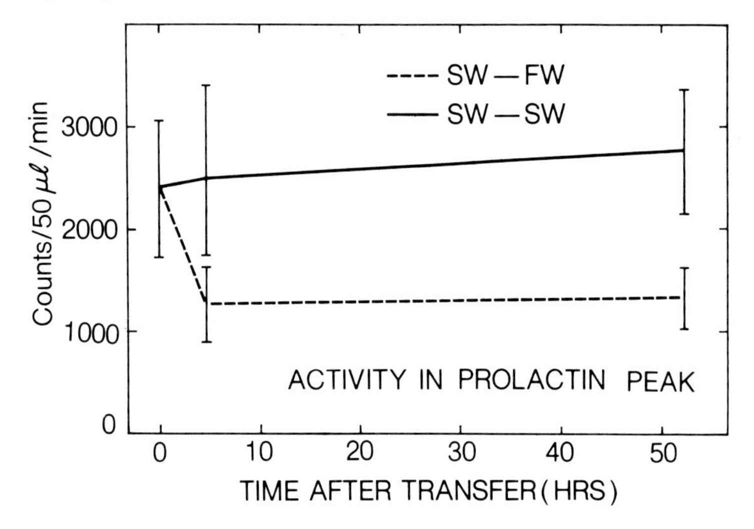

Source: from Brewer and McKeown, 1980

Figure 1.9: Cold-brands 48 hrs after marking juvenile coho and sockeye salmon.

Source: from Mighell, 1969

become much higher than for other marking techniques and large numbers of fish can be individually identified. These techniques are relatively easy and require little equipment. However, the type of equipment is often not readily available, especially for remote field applications. These techniques are fairly rapid to apply and marks are often clear and last for over a year. The pigment cells that are damaged during the branding process allow the mark to become visible but cells can often regenerate and cover the scar mark. Brands are used to make the fish highly visible and recognizable to a fish sampler but unfortunately the mark makes the fish more obvious to a predator and rates of survival will be different for these marked fish.

In order to avoid highly visible marks that might attract predators, a number of other more inconspicuous types of markers have been developed. One such approach has been to spray fish with a fluoresecent dye (Phinney *et al.*, 1967). The dye becomes embedded in the scales and is not visible under natural light conditions. However, when the fish is illuminated by ultraviolet radiation the dye fluoresces and becomes visible. Tests have indicated that there is no reduction in survival by the use of this technique. This method does have problems in that the dye marker becomes scattered in the body of the fish and as the fish grows the dye can be lost. Also, procedurally it may be inconvenient because of the necessity to provide ultraviolet illumination.

Tetracycline antibiotic is another substance that does not affect fish growth or survival and can be administered easily by injection, orally or simply by immersion (Weber and Ridgway, 1967). This dye is deposited specifically in areas being calcified such as vertebrae, opercular bones and otoliths (Figure 1.10). The tetracycline remains as a permanent marker and can be visualized as a fluorescent compound when exposed to ultraviolet illumination. If successive applications of tetracycline are applied to fish additional fluorescent bands can be laid down on bony elements. This allows for a large number of possible combinations to be attained as well as leaving a marker that can be identified with respect to time of deposition, so that information on rates of growth, especially of bone tissue, can be obtained. This method does, however, suffer from problems of applying uniform amounts to all individuals in a population, the bone tissue is more difficult to obtain for detection than a tissue from topically incorporated dye and ultraviolet illumination is again required.

Radioisotopes have also been employed as markers for fish (Carlson and Shealy, 1972). Many water-soluble radioisotopes are available and can be incorporated into fish tissues easily by immersing fish or by oral administration. Isotopes to be used must have high enough energies of emission for detection and possess appropriately long half-lives. Although these markers require a radiation detector they are useful for identifying particular marked fish as well as allowing identification of predators that have consumed the marked fish.

An interesting combination of radiation and X-ray fluorescence has been used to mark fish (see Calaprice *et al.*, 1971). When the electrons of different atoms, which are in various energy states, are exposed to radiation energy, changes take place such that electromagnetic radiation is released. X-rays produced in this way can be detected and their elemental origin can be determined. Because of the various elements found in fish tissues or in materials applied to fish, a marking technique becomes available in the form of a chemoprint. Calaprice and Calaprice (1970) have used a technique whereby desired oxides are mixed with silicone rubber and injected into fish tissue. The resulting hardened rubber becomes suspended as a small permanent innocuous elemental

Figure 1.10: Saggital otolith from a 30-day-old Chinook salmon alevin from Capilano River hatchery, British Columbia showing different primordia and daily growth rings (top). The lower figures are of a saggital otolith from a 130-day-old Chinook salmon fry. This fish received tetracycline treatment at an age of 116 days. The tetracycline leaves a lightly coloured ring that is visible in the figure on the right (arrow).

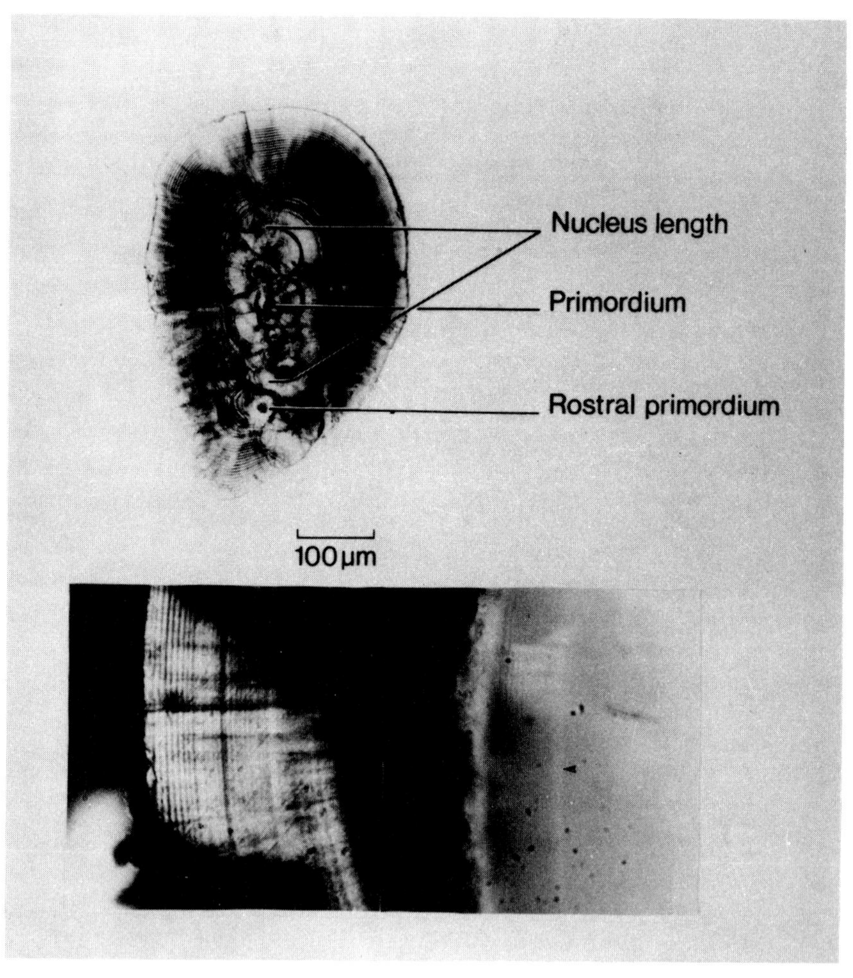

Source: from Nielson, 1983

marker. This technique can be long lasting and by the use of different elements a coding system can easily be established. The only obvious disadvantage is the type of equipment required for analysis. Calaprice (1971) has used this technique to analyse the X-ray spectrometric results obtained from untreated fish from different geographic regions. Fish living in different environments acquire a different chemical composition which is quantifiable by this technique, and the fish therefore possess their own natural tag. Mulligan *et al.* (1983) have recently employed X-ray spectroscopy to identify the specific elemental chemical composition of salmon vertebrae and were fairly successful in classifying different fish stocks on these composition characteristics (Figure 1.11).

Figure 1.11: X-ray spectrum of some bony elements from adult sockeye salmon optimized to observe the elements from vanadium through zirconium.

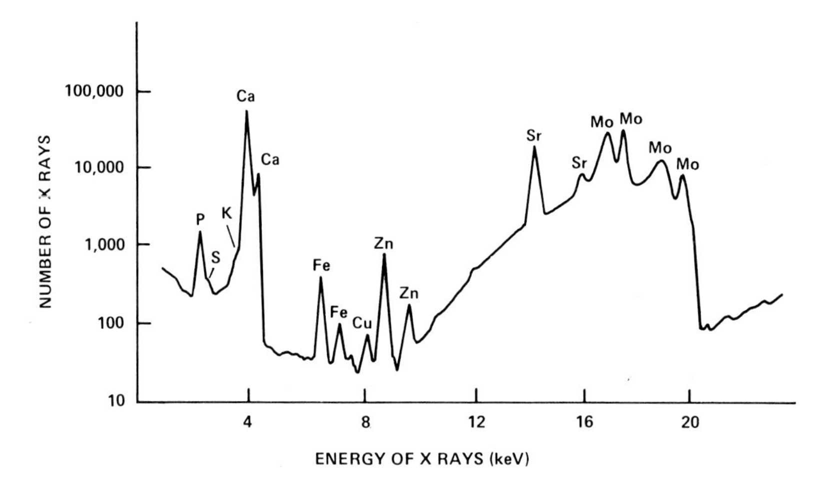

Source: redrawn from Mulligan *et al.*, 1983

Another type of natural marker that may be useful is the presence of specific types of parasites (see Harden-Jones, 1968). Fish might acquire a specific parasite early in life and only from a particular environment. If this occurs a natural marker exists that can be used later when fish of the same species, but of different stocks, mix at some later stage in their life cycle. Promising results have been obtained with some Pacific salmon stocks that acquire certain parasites early on in the freshwater phase of their life which are retained as identifying markers later when the fish are in the sea.

Instead of producing some mark on a fish for identification, a variety of tags have been developed to attach to fish for recognition (Beckett, 1973). The tags vary from small metal clips that can be pinched onto opercula, fins, jaws, etc. to tags such as Petersen discs which can be inserted through the dorsal musculature with discs or buttons attached to each side. Yet other tags, such as spaghetti tags, can be implanted into muscle tissue and have various attachments for identification. The disadvantages of these types of labels are mainly their effects on survival and loss of tags. These disadvantages become less important, however, if appropriate experiments are conducted so that mortalities and tag losses are known for any particular procedure. Almost all procedures for marking or tagging fish require catching, handling and releasing the fish. Depending on the procedures this in itself can cause stress to fish and cause death directly or indirectly by the fish being in a weakened condition and thus more susceptible to predation. Tagging of fish also causes additional stress due to the physical attachment of the label. After implantation of a tag the fish may consequently still suffer from the physical presence of the tag or from damage done or continuing to be done by the attachment. Tags are often good markers for short periods of time but they can always be lost due to being accidentally removed as time passes. Fish growth also causes tags to be displaced and lost.

Jefferts *et al.* (1963) have developed a type of tag that appears almost ideal as it is easy to implant, causes little mortality, usually stays with the fish for long periods of time and can be detected easily by researchers but not by the fish or its predators. This procedure entails the injection of a small (1 × 0.25 mm) piece of magnetized stainless steel wire into the nose of the fish (Figure 1.12). The wire can be color coded or marked with notches to provide

Figure 1.12: Tagging by injection of small (1 × 0.25 mm), coded, magnetized stainless steel wires into the nose region of a Pacific salmon. Since tag is not visible from the exterior, the adipose fin is clipped off for identification. The presence of the tag can later be checked by a magnetic detector and the individual fish can be later identified by its particular coded wire implant.

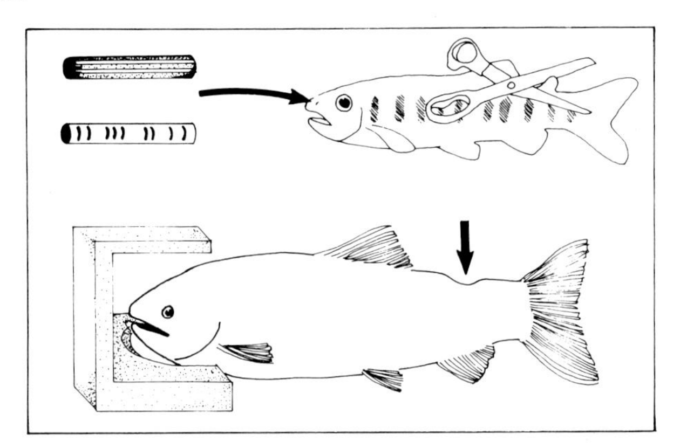

a binary code. This method is extensively used for tagging salmonid fishes. Since the tag can not be seen externally the adipose fin is removed to indicate the presence of the wire tag. The removal of the adipose fin does not result in any increase in mortality and has no other apparent effect on the fish. When a fish that is missing an adipose fin is recovered it can be easily checked for the presence of the coded wire with a magnetic detector. Durkin *et al.* (1969) have also devised a method that can fairly successfully separate out wire-tagged migrating salmon.

Direct Observations

Fish have been observed from land, ships, aircraft and under water. Although of limited use, some information can be obtained by direct observation on the direction, speed of travel, obstacles and numbers of individuals migrating. More useful information has been gathered about fish movements by attaching devices which can be detected without the necessity of visual contact with the fish. Hasler *et al.* (1958) used a simple but effective device to follow white bass, *Roccus chrysops*, to their spawning area after being displaced in the lake. The studies were designed to elucidate the environmental cues used by the fish to orient to their final goal. Fish were tracked by following a small float attached by nylon line to the fish (Figure 1.13).

Figure 1.13: Small float attached to the dorsal fin of a white bass by a nylon line for direct surface observations of fish movements.

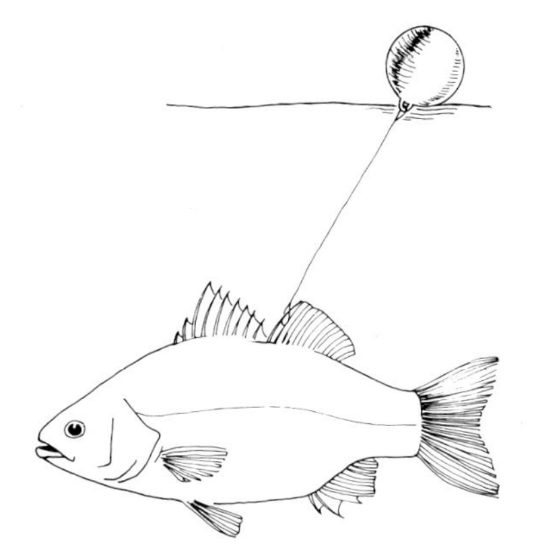

Source: redrawn from Hasler *et al.* 1958

These techniques have been replaced by others such as echo-ranging or sonar, in which sound travels through water at a given speed and thus distances can be calculated. The sound waves can be transmitted vertically as in echo-sounding (depth) or can be transmitted horizontally as in an asdic (distance and direction). When the sound is reflected from fish or other objects problems of identification arise. Many of these problems have been worked out for different sizes of fish and at different depths. For example the International Pacific Salmon Fisheries Commission uses such a technique to assess the numbers of sockeye salmon smolts migrating each spring out of Cultus Lake, British Columbia. Chart recordings of echo-sounders are used to follow the numbers of fish as they move up from the bottom at night and begin their movement towards the outlet stream.

In this case fish are simply reflectors of sound, but techniques have been developed whereby the fish are fitted with a sound transmitter which can be followed with an appropriate receiver (Figure 1.14). The ultrasonic transmitters are such that they can be implanted into the fish and last for a considerable time. With such a transmitter a single fish of known species, size, age, sex, etc. can be tracked for long periods and with greater precision than alternative

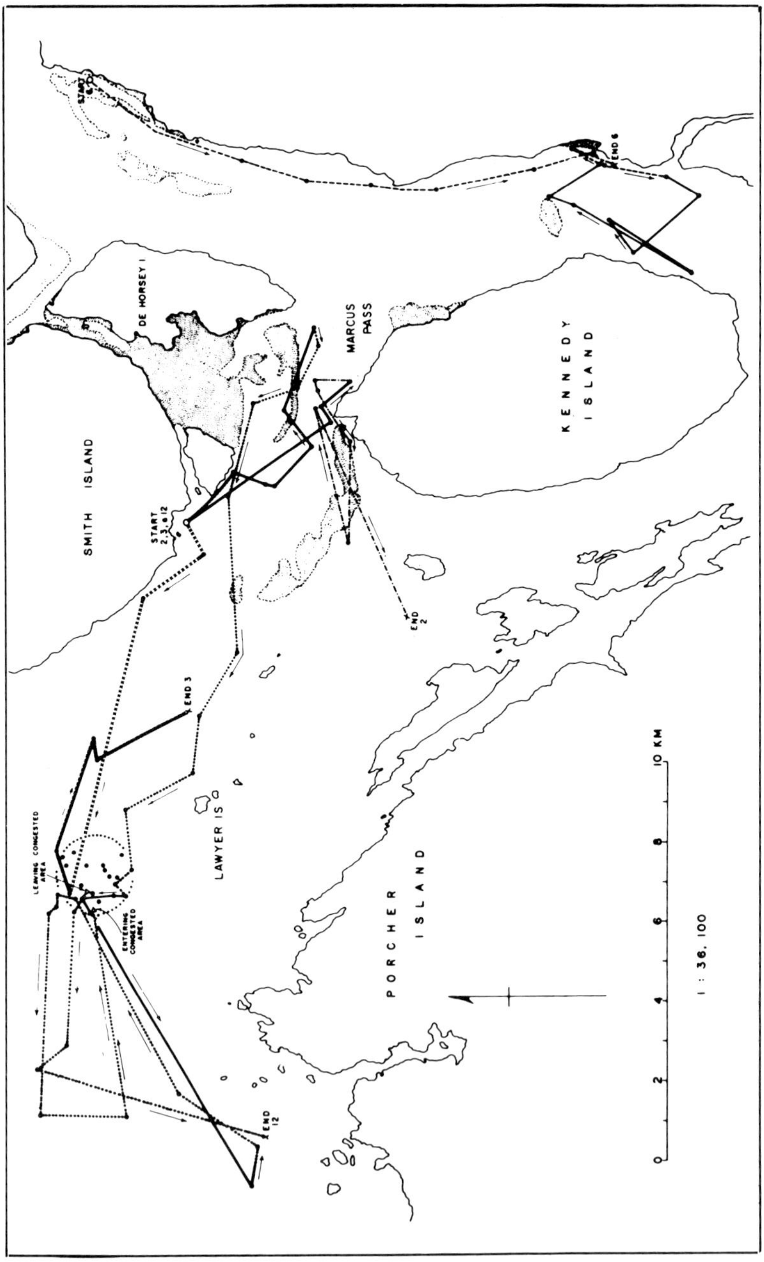

Figure 1.14: Ultrasonic tracks of four adult migratory sockeye salmon entering the Skeena River estuary, British Columbia.

Source: from Groot *et al*. 1975

techniques allow. Furthermore, external transmitters which are sensitive to temperature or salinity provide valuable additional information on the environment through which the fish passes.

Identification Characteristics

In studies on migration it is often necessary to distinguish one population of fish from another. Identification may be required for studies on evolutionary development of specific migratory populations or may become important for determining the migratory patterns of individual populations which become mixed during some periods in their life cycle. Yet some populations resemble each other so closely that even morphological criteria such as number of gill rakers, fin rays, pyloric caecae, etc. cannot be used for identification. Fortunately, structures such as scales and ear bones possess features which are characteristic of the environments to which the fish has been exposed. These characteristics can be extremely useful in migration studies since they can provide detailed information on the life history of the fish. Even in instances where it is not possible to correlate scale or earbone features with known life history, it may still be possible to distinguish different populations using such features.

Scales are particularly important for ageing, and for identifying stocks of many species. Soon after scales first develop they start to grow out from a central area called the focus, centrum or nucleus (Figure 1.15). As a scale grows

Figure 1.15: Scale from a maturing chum salmon showing individual circuli (A) which come close together during slow growth periods and become discernible then as an annual ring (R).

Source: from Lalanne and Safsten, 1969

it displays concentric ridges called circuli on the upper bony layer. Sometimes these circuli are only present clearly on the anterior portion of the scale or only on specific scales on an individual fish. The circuli on some fish such as herring are not circular but run across the anterior portion of the scale. When growth is rapid the circuli are far apart; when growth is slow such as in winter the circuli are close together and can be detected as an annulus. Since the growth of circuli can be affected by such things as temperature and food supply, the habitat of the fish can often be reflected in the pattern of circuli which are therefore useful as stock identifiers.

Like scales, certain bony elements of the inner ear of bony fishes are useful for ageing and identifying populations (Figure 1.10). The inner ear responsible for hearing has three pairs of chambers which are similar in structure. Within each utriculus, sacculus and lagena there is a bed of sensory hairs in close contact with a calcareous structure, the otolith. The otoliths in the utriculus and sacculus appear very early on in development and continue to grow throughout life. As an otolith grows layers are deposited in the form of alternating dark hyaline and opaque bands. The differences between the bands may be due to differences in calcium deposition brought about by changes in temperature or food supply. The consequences, however, are that annual layers may be observed for identification purposes. Pannella (1971) has also described daily growth increments in fish otoliths. These daily changes in otolith microstructure permit improved resolution of age as well as providing refined estimates of fish growth. Subsequent work has revealed that both increment number and width could be affected by such parameters as temperature, photoperiod and feeding frequency (see Neilson and Geen, 1982). Since the daily increments are affected by environmental parameters, they may provide a detailed life history and migratory sequence of various fish populations.

Different fish populations have been shown to exhibit various distinct morphological characteristics. Many of these populations also display distinct biochemical characteristics useful for identification purposes. Variation in the structure of different compounds may be brought about by environmental as well as genetic factors. Regardless of the causes of the observed differences, such variations are still useful for identification. Thus, differences in fish populations due to variations in blood and muscle proteins have been described. Various enzymes differ slightly in their amino acid composition between populations and their physio-chemical characteristics may thus provide useful markers.

Laboratory Techniques

There have been many technological advances over the past few years to provide a number of biochemical assays and operational techniques useful for the study of fish migration. To list and discuss these procedures here would go beyond the scope of this chapter. The reader is thus directed towards the appropriate sources for detailed procedures. In general, however, new sampling techniques and assay procedures have allowed for the quantification of many different types of organic and inorganic compounds and at very low concentrations. For the physiologist this has opened the door to the possibility of answering many new and exciting questions.

2 PATTERNS OF MIGRATION

Introduction

During the life cycle of many fish species there is a period that can be delineated as a migration. The form or pattern that the migration takes, however, varies considerably between different fish species and even within species. Thus, the distance covered during a migration may vary from short daily displacements for food gathering purposes to extensive journeys of hundreds of kilometers such as the migrations of many Pacific salmon. Migrations that cover similar distances may vary considerably with respect to duration. For example, some fish feed along the way whereas others must rely upon stored body reserves. The duration of a migration may also depend on the speed of movement which in turn largely depends on many energetic, physiological and morphological characteristics inherent in any particular species.

Migratory behavior is associated with appropriate orientation mechanisms in order for a fish to be able to find its way from place to place (see Chapter 3). The particular pattern of the migration may depend on what cue is being used for orientation. Fish are known to be able to change from one orientation mechanism to another depending on the prevailing environmental conditions. Also, during different phases of a migration, an individual fish may use different cues and at some point may switch to some different mechanism. Thus, adult Pacific salmon may use current, celestial or magnetic cues while migrating in the open ocean but may change to olfactory stimuli upon entry into a home stream environment. Recent ultrasonic tracking studies indicate that many fish do not swim in a direct line to the final goal but display many wandering movements or side trips. These hourly or daily deviations in migratory patterns may be a result of different orientation mechanisms coming into play or a search being carried out for one particular stimulus.

Many fish, whether they be long or short distance migrants, return to areas that were the original starting points for the migration. Other species of fish, however, migrate from one place to another and may never return to the original starting point. This latter pattern may be termed a one-way or a non-return migration. Particular return migrations also vary in that the start and finish places of the migration are the only points in common. After initiating such a migration, the fish may simply wander from place to place, never retracing its track until arriving back at the starting point. This type of movement may be termed a loop or circular migration. On the other hand, some fish species return to the starting point by retracing the outward bound journey. Baker (1978) terms this type of movement as a to-and-fro migration. Some return migrations may be combinations of circular and to-and-fro migrations. Thus, certain individual races of Pacific salmon may descend a natal stream, migrate in a circular fashion in the ocean and then as adults traverse the home stream in an opposite direction. These patterns of migration are exemplified in Figure 1.2.

The migratory pattern of most species of fish shows periodicity, although some are irregular in their movement patterns. Regular migratory movements may take place daily, monthly, annually or over some other period of time depending on the species and the stage of the life cycle. For example, young sockeye salmon, *Oncorhynchus nerka*, smolts display vertical daily migrations in lakes and annually migrate to the ocean or return to the spawning grounds. As Baker (1978) points out, the periodicity of migratory behavior differs between the individual and the species in general. Annually, there are seaward migrations of sockeye salmon smolts to the ocean but there is a small proportion of the population that stays in the lake for another year. Thus, the pattern of migration for this species is different for various individuals. Likewise, spawning adult sockeye salmon return every year but some precocious individual males or jackes return one year earlier than most of the population.

The migratory patterns differ both within and between fish species due to a number of different biotic and abiotic factors. Many fish migrations take place in order to locate a suitable spawning area. Each species will obviously require different environmental characteristics that determine the pattern of migration for that species, but from year to year some of these characteristics may change and thus also will change the migration route or destination. Other migrations are for the purpose of obtaining food or avoiding predation. Prey and predator numbers change and, in accordance, so will the movement of fish associated with such population fluctuations. Some fish migrate seasonally to avoid harsh environmental conditions or take advantage of favourable climatic conditions. These annually changing events can cause changes in migration initiation, termination or patterns followed by the fish migrants.

Even though migration patterns differ interspecifically and from year to year intraspecifically, variations also exist between different races of the same species. Sockeye salmon characteristically emerge from the gravel and migrate to a nursery lake for one to two years, migrate to the ocean for two to three years and finally return to spawn in their natal stream. However, kokanee salmon, a freshwater race of sockeye salmon, do not migrate to the sea and spend their entire life in fresh water even though they have the opportunity to migrate to sea. Other Pacific salmon species transplanted to the Great Lakes System of North America have also adapted to freshwater migratory routes throughout their entire life cycle. Other Salmonidae such as certain species of trout, char and whitefish also have some races which are entirely freshwater and others which migrate to the sea. Similar diverse migratory routes may be seen in individuals of numerous families of teleosts such as smelts (Osmeride), sticklebacks (Gasterosteidae) and alewives (Clupeidae), to mention only a few.

Since migratory patterns differ between species as well as within species it is diffiult to categorize the various basic types of fish migrations. However, in order to discuss the array of migratory patterns exhibited by fish, the terminology and outline of fish migrations by Myers (1949) will be used. *Diadromous* fish are those that migrate from fresh water to sea water or the reverse. More specifically, fish that spend a large proportion of their life cycle in the ocean and return to fresh water to breed are termed *anadromous*. Examples of anadromous species are certain salmon, trout, char, smelt, whitefish, alewives, shad, sticklebacks and lampreys. Diadromous fish that spend most of their lives in fresh water but migrate to the sea to reproduce are termed *catadromous*. The best known catadromous fish are the freshwater eels (Anguillidae). Fish from the southern hemisphere such as some of the southern trout (*Galaxias* spp.) as well as certain smelt and grayling species also display catadromous behavior. Other examples are the Japanese ayu, *Plecoglossus altivalis*, and the Hawaiian climbing goby, *Awaous guamensis*. Myers (1949) describes a third type of diadromous fish termed *amphidromous* that refers to fish that can migrate from fresh water to the sea or vice versa. However, this type of diadromous migration is not for the purpose of breeding but occurs regularly at other times during the life cycle of the fish. Migrations of some gobiid fishes may be of this type. Some fish species spend their entire lives in fresh water but still have well defined migration routes within lakes, streams and rivers and are classified as *potamodromous* fish. In contrast to freshwater species, other fish live and migrate entirely within the sea water environment and are termed *oceanodromous*. Migrations of cod, herring, plaice and tuna are well documented cases of such ocean migrations.

Baker (1978) devotes special attention to animals that migrate within the shallow littoral zone. Many fish migrate within this zone in accord with daily or tidal cycles. This zone is influenced by changes in temperature, light intensity, salinity, day-length and other physical as well as biological factors and thus provides variable environmental stimuli causing regular migratory behavior. Certain blennies and mud skippers display migratory patterns in accordance with regular changes within this littoral zone. Other fish species that migrate within the littoral zone display regular patterns as well but the frequency is reduced. The grunion, *Leuresthes tenuis*, displays an interesting species-specific migratory pattern during the breeding season that is in tune with high tides. Sexually mature adults swim in with waves during night time at high tide a few days following a full or new moon. The female digs herself into the sand and the males surround her to fertilize the eggs. Two weeks later the eggs are washed out of the sand and hatch with the next lunar cycle of high tides. This type of specialized migration pattern has probably developed in order to reduce predation on the developing offspring.

Harden-Jones (1968) and Baker (1978) have presented excellent reviews of the migratory paterns of a number of different fish species. Only a few will be considered here as examples of the types of migration that fish undertake and the environment through which they must travel. Only well documented migrations within the ocean, fresh water or between the two environments will be discussed here. However, the diversity of fish migratory patterns far exceeds these examples and specifics concerning individual species should be sought in appropriate literature sources.

Oceanodromous Migrations

The oceans of the world support a vast number of different fish species which are distributed over the major portion of the globe. If the vertical as well as the horizontal distribution is considered, the size and scope of fish migratory routes becomes large indeed. Many fish species only undertake short exploratory movements throughout their life and others may display short but regular inshore/offshore return migrations. Yet other fish species undertake extensive long-distance migrations such as those of herring, cod, tuna and plaice. This latter type of movement will be considered in this section. Considering the size and conditions of the environment through which these migrations take place, it is not surprising that there are large gaps in our understanding of these fish movements.

The long-distance migrations of oceanic fish species have been described as a circuit rather than a back and forth type of migration. The circuit may take all sorts of shapes and may encompass both the horizontal and vertical planes. As outlined by Baker (1978), these migratory patterns must include specific areas that meet the required demands of the migrant at precise times in the life cycle of the fish. These include a spawning area, nursery area, feeding area, winter area, summer area, migratory area, etc. These different areas are visited in sequence by an individual fish and between species or demes of the same species, and each area may overlap to different degrees both in respect of time and space (Figure 2.1). Migratory fish must be capable of arriving at these different areas at the appropriate time and stay long enough to extract the requirements from that particular environment. Each area has different physical, chemical and/or biological characteristics and the fish is likely to be capable of detecting these differences. Unfortunately, little is known about how fish distinguish these areas.

In most areas of the ocean there are water currents, which may flow in a circular (gyral) or in a counter-current manner. In the latter case two water masses, which are vertically or horizontally adjacent, flow in opposite directions. Fish may return to a starting point by swimming in the counter-flowing current with obvious savings in energy expenditure. Baker (1978) points out that if particular demes (local populations) of fish could exploit these ocean currents, they could move largely by means of drift but alter the timing of the migration circuit by depth regulation, changing swimming speed, varying swimming direction or amount of time spent in station-keeping.

In order to arrive at a suitable area at an appropriate time in the life cycle, the oceanodromous fish must be able to recognize the area by characteristic environmental stimuli or possess some type of biological clock that is capable of timing

Figure 2.1: Patterns of movement in oceanodromous fish migrations. Adult fish may move in different directions on the diagram depending on what time of year they spawn. The juveniles may move from nursery to feeding and winter areas and they may be recruited into the adult population at different times from different areas depending on the species or populations.

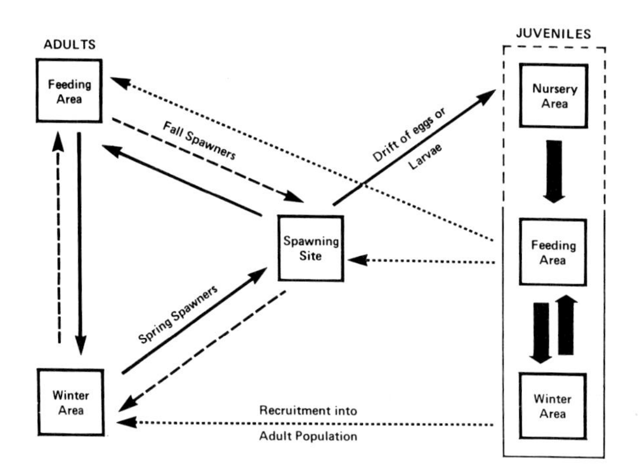

Source: redrawn from Dingle, 1980

the different aspects of the migration circuit. The endogenous clock may simply be the attainment of an appropriate physiological state coinciding with arrival at a particular area. Thus, gonadal maturation might be the sign stimulus for a migrant indicating arrival at a spawning area. If the migrating fish recognizes a suitable area by environmental stimuli, it may be because the fish can detect some aspect in the environment that identified that area as suitable to the fish for eliciting the desirable behavior at that particular time in its life cycle. On the other hand, the migrating fish may recognize the area because it had been there before and had learned or imprinted to some environmental cue.

The oceanic migrant may recognize a particular area such as a natal spawning site, by olfactory discrimination in much the same fashion as a migrating salmon might orient towards the home stream in order to locate its natal spawning ground. If marine fish migrate with oceanic currents, olfactory cues might be present that have emanated from the spawning area. These migrating fish could easily return to this spawning area by staying in sensory contact with a gyral current or by swimming upcurrent to where the olfactory stimulus originated. The latter type of migratory pattern for a fish following an olfactory cue up-current would be enrgetically costly. However, if the fish were in a counter-current system it could displace itself into an adjacent current flowing back to the spawning area and occasionally return to the outflowing current to reorient to the olfactory stimulus.

Most oceanodromous fish that have been extensively studied thus far appear to migrate with ocean currents (see Harden-Jones, 1968; Baker, 1978). From initial observations, however, it appears that some fish migrate against the ocean currents. Certain demes of Atlantic herring, *Clupea harengus*, return to the Norwegian coast across the Norwegian Sea which is a direction against the Atlantic drift. However, these fish may avoid this current by swimming at greater depths, possibly in currents flowing in an opposite direction. The return migrations of the tunny, *Thunnus thynnus*, that spawn in the Mediterranean Sea occur at greater depths than the outward journeys and may indicate fish seeking out appropriate currents to reduce migration cost.

Herring

The herring are found in both the North Pacific and North Atlantic Oceans (Figure 2.2). Depending on the author, both Pacific and Atlantic herring have

Figure 2.2: The geographical distribution of the Atlantic and Pacific herring.

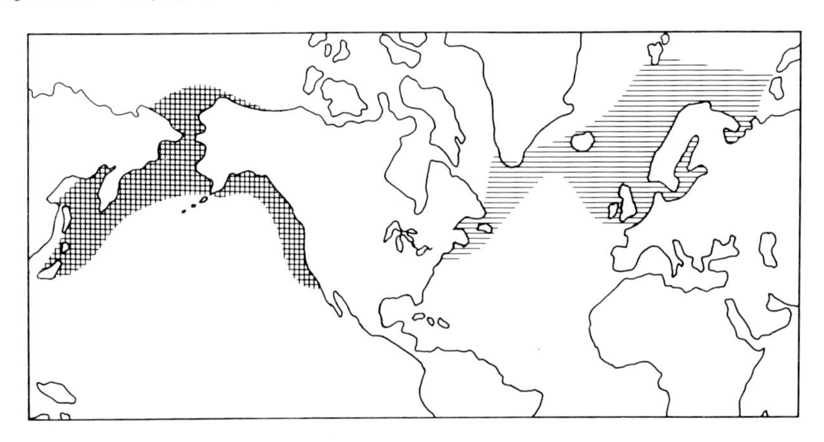

Source: redrawn from Baker, 1978

been grouped together as a single species or the Pacific herring has been classed as a different species, *Clupea pallasii*. Anatomically, the Pacific and Atlantic herring appear very similar, with only slight differences evident in the ventral scutes or scales on the underside of the body anterior to the pelvic fin. Both types of herring have similar feeding habits, produce demersal eggs and have pelagic larvae. Atlantic herring spawn at various times throughout the year and do so at depths of 40-200 m. They spawn on shingle or gravel beds which may sometimes be far from the shore or oceanic rises. Pacific herring spawn in the spring with heaviest concentrations in March. They spawn in shallow coastal waters from high tide to 11 m deep and deposit their eggs on various seaweeds as well as different abiotic substrates.

Herring display seasonal migrations that cover a relatively large area. They also perform daily vertical migrations over short distances and for different purposes than the longer seasonal migrations. Observations of Atlantic herring stocks indicate that these fish inhabit deep water during the daytime and migrate to the surface at sunset. At midnight they sink, only to rise once again near dawn before retreating to greater depths for the rest of the day. The zooplankton upon which the herring feed also display diurnal vertical migrations and the herring daily migration cycles may in part be due to the movement patterns of their food source. The herring may exhibit these daily movement patterns to reduce the risk of predation by avoiding the brighter upper water layers during the day. Alternately, or in addition to predator avoidance, McLaren (1963) suggested that the reduced metabolic rate of fish in the cooler, deeper water where vertical migrants move when not feeding could leave more of the ingested food energy available for growth and reproduction.

More information is available on the migratory patterns of Atlantic herring and thus this group will be considered here. The Atlantic herring are subdivided into a number of different groups, races or demes and each group has further subdivisions. The Atlantic herring are divided generally into a northwest and northeast Atlantic population. The northeast Atlantic group can be subdivided further into seven different groups that do not intermingle to any great extent. These include the Atlanto-Scandian, Hiberno-Caledonian, English Channel, North Sea, Skaggerak, Baltic and isolated stocks in certain fiords and estuaries (Figure 2.3). Within these groups there are still smaller subdivisions of herring races that spawn in the same area at the same time of year and follow the same migration circuit. There is more intermingling between individuals of these smaller races but the majority of each population remains distinct. The four groups that Baker (1978) summarizes will be outlined here as an example of herring migration circuits and of how they overlap spacially and temporally. One is the Norwegian spring-spawning group which belongs to the larger Atlanto-Scandian group. The other three belong to the North Sea group and are the northern North Sea summer-spawning group, central North Sea autumn-spawning group and the Southern Bight winter-spawning group.

The annual migratory pattern of the Norwegian spring-spawning group is shown in Figure 2.4. This group of herring arrives off the Norwegian coast in the early spring from a westerly direction. The fish then move close to the coast near the spawning grounds about five to six weeks before spawning. Spawning occurs between February and April after which the herring appear to move north, northwest and west. These fish may be swimming with the western branches of the Atlantic current which leads them into the gyral current of the Norwegian Sea. They move north with the current during spring and summer at a time when populations of copepods and euphausids are increasing. The herring may thus be following currents and a summer food production. During this time the fish are at depths of 30-40 m, they do not exhibit daily vertical migrations but remain in the warmer Atlantic water that overlies the colder Arctic water. In the fall these herring move south in the direction of the East Icelandic current. At this phase of the migration the fish start daily vertical migrations

Figure 2.3: The distribution of the six main groups of herring in the northeast Atlantic Ocean (Atlanto-Scandian, Hiberno-Caledonian, English Channel, North Sea, Skagerrak and Baltic). There are also more or less isolated stocks in certain inlets and estuaries. Although there is some interchange between the groups, individuals from one group do not mix freely with those of other groups. Within one group there may be several spawning stocks.

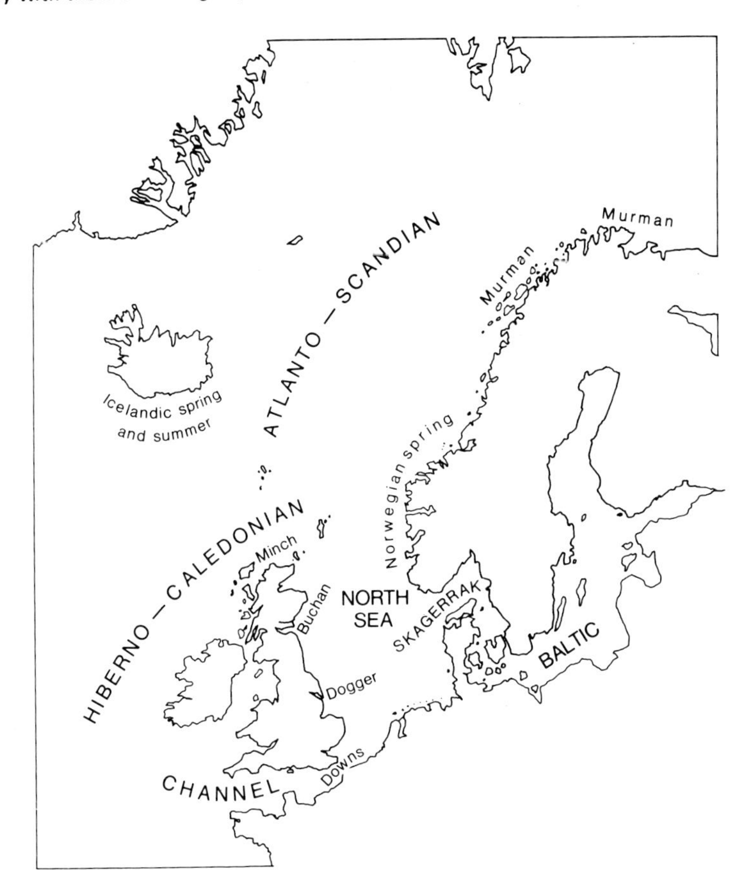

Source: redrawn from Harden-Jones, 1968

from 300-400 m during the day up to the surface to feed during the night. Prior to completing the circuit off the Norwegian coast, this group of herring crosses the northeast flowing North Atlantic current. This movement might be achieved by the herring remaining in the East Icelandic current which flows under the North Atlantic current towards the spawning grounds.

Following spawning the resultant larvae are carried by the prevailing currents from the spawning grounds. Subsequently, they do not enter the migration circuit described above but instead move to the fiords and coastal waters of Norway where they remain until they increase in size. During these earlier years

Figure 2.4: The winter, spring, summer and autumn distribution of adult herring of the Norwegian spring-spawning stock that is part of the Atlanto-Scandian group. The solid black areas in the top-left diagram indicate the spawning areas. The horizontal shading shows the approximate distribution of the herring during the season indicated and the dashed line depicts the distribution of herring during the previous season. Surface water currents are indicated by arrows.

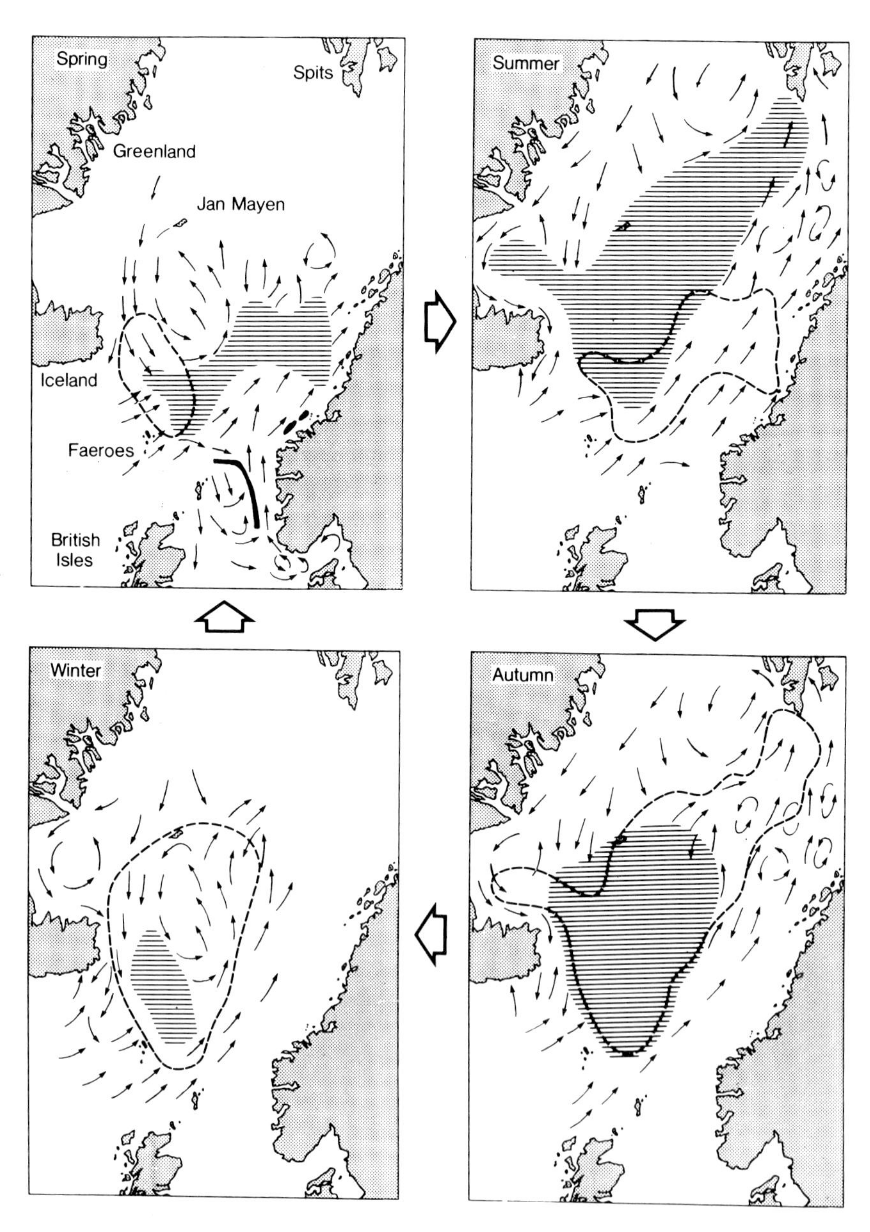

Source: redrawn from Baker, 1968

the herring perform annual littoral migrations whereby they inhabit inshore waters during the summer and offshore waters during the winter. Once the young herring have reached a large enough size they join older fish in the migration circuit and remain there until they return to the spawning grounds when sexually mature. The time of the return varies from the fourth to the seventh year of life. Evidence suggests that the young herring enter and complete the same migration circuit as their parents and that they return as sexually mature adults to their natal area.

The migration circuits within the North Sea of three races of herring are depicted in Figure 2.5. These three migration circuits overlap both in time and space and all appear to feed in a similar area in the northern part of the North Sea. Evidence suggests that some fish migrate from one group to another but the majority spawn with the group they joined when mature. Also, most of these herring seem to recruit to the group that spawned in their natal area. Following hatching the larvae are carried by ocean currents and subsequently spend their early life in inshore waters around Germany, Denmark and Scotland. In the spring of their first or second year when the herring are approximately 10 cm in length, they migrate to deeper waters particularly south and east of Dogger Bank in the southern North Sea. These fish move west, east and north in the fall and even further north in the following spring and summer. The sexually immature fish that do not spawn feed in the west and northwest part of the North Sea and winter further to the east. Most of the adult migrations appear to move with the prevailing ocean currents. A few groups, however, seem to move against ocean currents during part of their life cycle, but until more detailed accounts of the duration and of the depth at which these fish swim are available, the actual orientation to currents will not be completely understood.

Cod

The cod, *Gadus morhua*, has been a major protein source for a large number of people over at least the past millennium. This species has a wide distribution over the North Atlantic Ocean, ranging from the middle United States to Baffin Island on the west, to Northern Europe, the North Sea, the Baltic Sea and as far east and north as the Kara Sea (Figure 2.6). Within this total population there are several distinct groups that do not appear to intermingle. There are thus populations found around Newfoundland, The Faeroes, in the North Sea, the Baltic Sea, along the Norwegian Coast and adjacent Barents Sea, and between Iceland and Greenland.

Cod usually spawn from February to June in cold water near the bottom or further up in the water column. Females spawn over a period of several days and can produce over 10^7 eggs. Fertilized eggs are distributed in the upper 100 m of the ocean and hatch in approximately two weeks. Up to six months following hatching the young cod are found in midwater feeding on zooplankton, particularly copepods. They then descend to the sea floor where they feed on benthic organisms such as certain crustaceans. During this stage they may

Figure 2.5: The annual migration circuits of three different spawning stocks of herring from the North Sea group. The migration circuits of these three stocks overlap in both space and time. (a) The North Sea summer-spawning stock. (b) The central North Sea autumn-spawning stock. (c) Southern Bight winter-spawning stock. The solid black indicates the spawning areas and the horizontal lines indicate the wintering areas. The dotted lines indicate the position of the Dogger Bank. The arrows indicate the likely migration routes (long) and the surface ocean currents (short).

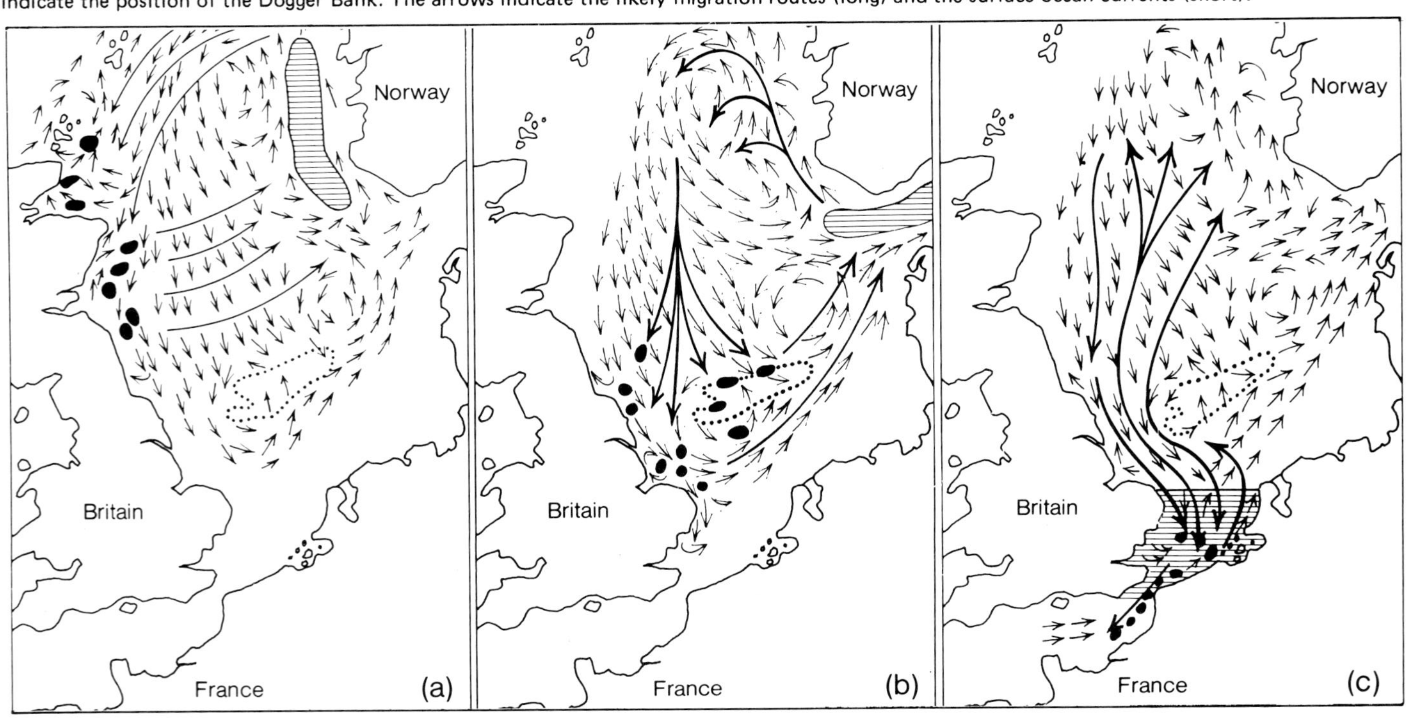

Source: redrawn from Baker, 1978

Figure 2.6: The distribution (dotted lines) of three main groups of Atlantic cod (Newfoundland, North Sea and Baltic, Barents Sea and Norwegian coast). The solid black shading indicates the major spawning areas whereas the stippling indicates minor spawning areas. The horizontal lines indicate wintering areas. The arrows indicate adult migrations (continuous) and drift migrations of eggs and larvae to nursery grounds (dashed).

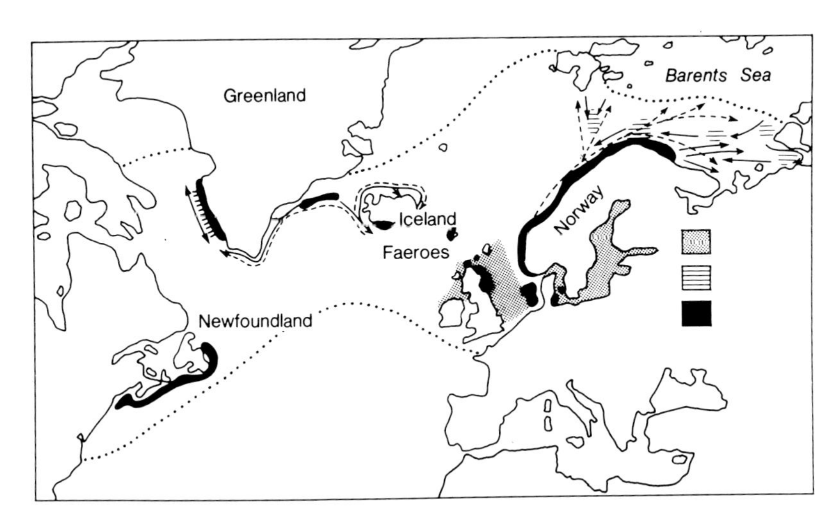

Source: redrawn from Baker, 1978

possibly migrate to shallow inshore waters during the summer and return to deeper waters in the winter much in the same way as do young herring. As the cod grow they change their diet to small pelagic fish and some crustaceans. At this stage they may then join mature fish of the same race and start a migration circuit. These fish may complete an entire circuit before spawning. Depending on the race of cod, spawning may occur at age 5 or be delayed for another 10 years. Having spawned once they continue to respawn annually for many years.

There are spawning populations of cod on both the west and east sides of Greenland as well as near southern Iceland (Figure 2.6). There appears to be an intermingling of these groups (see Harden-Jones, 1968). The west coast of Greenland supports a number of groups that spawn in the coastal inlets and make short migration circuits within fiords or out into coastal waters. Early in this century a warming trend in the north Atlantic resulted in a northerly extension of the distribution of many marine organisms. Increases in the cod populations on the west coast of Greenland are but one example. This increase in numbers of cod on the west coast of Greenland was probably due to the arrival of mature fish from Iceland.

The Irminger current flows from Iceland to southwest Greenland and is known to carry herring eggs and larvae from the Icelandic spawning grounds. Most of the mature fish resulting from this drift return to their natal areas. Some eggs and larvae originating from the southwest Iceland spawning-ground drift around to the north shores of Iceland and return as adults to spawn in their natal grounds. Thus, progeny from adults spawning off the southwest shores of Iceland can drift to different locations and complete different migration circuits before spawning. However, there is evidence that some fish can complete one circuit and then make another different circuit during a second migration. Adult cod on the west coast of Greenland include fish migrating in and out of coastal inlets as well as those coming from or returning to Iceland. There is a third group of fish in this area that has appeared over the last 20 years that has a migration circuit in a north-south direction along the west coast of Greenland. Although this may be a race of cod undetected earlier, it may also be a group of fish that has recently developed a new migration circuit and may thus be a useful source of information concerning factors that cause migration patterns to develop.

Cod that spawn along the Norwegian coast migrate north in April and spend the summer in the north and east parts of the Barents Sea. In the fall these fish move south and west to winter off the coast of Norway and eventually spawn near the coast in early spring. The young of these fish drift to the nursery areas which are in the northern part of the Barents Sea. The recorded returns to the feeding area in the Barents Sea are poor and may be only a reflection of the large size of the area. However, they may indicate movement between isolated races within the general area.

The eggs and larvae of cod from different races all appear to drift to their respective nursery areas but once the larvae sink to the bottom where feeding continues, energy must be expended in maintaining position against the current. When the young cod start to feed on pelagic fish, they follow their food source and migrate principally with the current. When cod migrate to the spawning ground they are generally in deeper water, whereas the spent fish seem to migrate in shallower waters with surface currents. The return spawning migrations may thus be in deeper counter-current systems.

Plaice

The plaice, *Pleuronectes platessa*, is a right-eyed flatfish or Pleuronectid that inhabits most of the shallow coastal waters of Northern Europe (Figure 2.7). It is found as far north as Iceland and The Faeroes and south along the coasts of France, Spain and Portugal with specimens being recorded in the Mediterranean Sea. They are found in the Irish Sea in the west and extend east to the North Sea, Baltic Sea and as far as the White Sea off the northwest coast of USSR. There appear to be distinct stocks of plaice near the Murman coast, Iceland, The Faeroes and in the Baltic Sea. Other races are present in the waters around Great Britain but movements between groups are more frequent. The Irish Sea plaice are somewhat isolated but some movement does occur through the English Channel into the North Sea. The migration patterns of the North Sea plaice are better understood than for other stocks and will thus be outlined here. Within the North Sea group there are four subgroups, represented by the

Figure 2.7: Geographical distribution of plaice, *Pleuronectes platessa*.

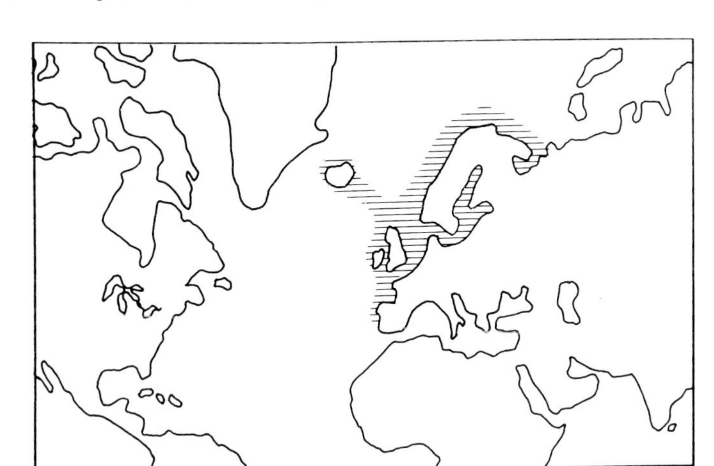

Source: redrawn from Baker, 1978

Scottish east coast, Flamborough, Southern Bight and German Bight stocks (Figure 2.8).

The eggs and larvae of plaice are pelagic and thus are carried by currents from the spawning grounds to nursery areas. Plaice have completed metamorphosis by about eight weeks of life when they descend to the bottom to feed mainly on benthic invertebrates. These nursery areas are illustrated in Figure 2.8 and are situated close to the coast due to the current patterns affecting the planktonic stages. During the first few years of life the plaice are found in relatively shallow water (up to 20 m) and as growth continues these older fish frequent deeper water. At sexual maturity which takes place at age 3-4 the plaice appears to return to its natal area. Following the breeding season from January to March, first-time spawners do not return to the deep water nursery areas from whence they came but disperse over larger feeding areas that overlap with those of other North Sea plaice races (Figure 2.8). Tagged spawning plaice exhibit a high degree of return to the same spawning ground and thus migration circuits might be repeated several times since the fish live to an age of 20 or more years.

Tuna

The tunas or tunnies are members of the Family Scombridae and possess remarkable physiological adaptations for fast continuous swimming during their extensive migrations. The group is distributed widely in temperate and tropical marine environments. Tunas are built for speed and not only have a streamlined form but also can retract their dorsal and pectoral fins to reduce drag. Tunas have the ability to deposit fat which is used during subsequent long distance

Figure 2.8: The distribution of three spawning stocks of plaice in the North Sea. The solid black dots represent the three major spawning areas. The one on the left is near Flamborough Head where plaice spawn in March. The bottom dot is the central Southern Bight where spawning occurs mainly in January. The dot on the right is the German Bight where spawning is in February. The fish from the Flamborough spawning ground have been recaptured out to the limits of the solid line, the Southern Bight fish to the broken line and the German Bight fish to the open circles. The three hatched areas indicate nursery areas and the arrows represent ocean currents.

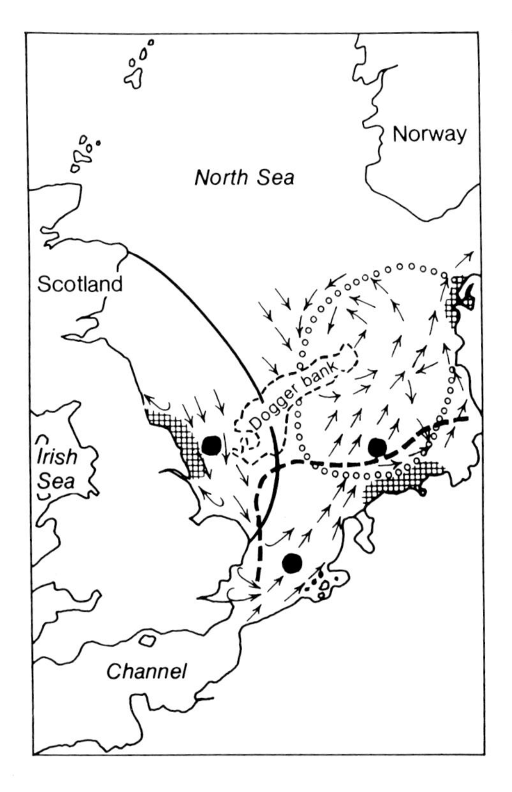

Source: redrawn from Baker, 1978

migrations both as an energy source and as a drag-reducing feature. The lipids increase the buoyancy of the tuna and deposits between connective tissue layers underlying the skin serve to dampen turbulence as water flows over the surface of the fish thereby reducing surface friction. Tuna can provide more metabolic energy for their muscles by saving energy from gill ventilation. They swim with open mouths so that water is forced across the gills – a process known as ram ventilation. Tunas must continually swim, however, in order to irrigate the gills. Tuna also have a larger gill area in comparison to body weight than most other fish. These fish also have a very muscular heart, a large blood volume, high hemoglobin concentrations and deliver blood to the tissues under high pressure, all of which allow for high performance of muscle tissue for swimming at speed.

Unlike most fish, tuna possess counter-current temperature exchanges between the anterial and venous sytem that allow them to maintain a body temperature above that of the ambient environment. This increase in body temperature may allow some tuna to feed in colder northern waters. A number of functions such as nerve transmission, muscular power output, digestion and assimilation will be increased with such a rise in body temperature.

A description of migratory patterns of fish would not be complete unless the remarkable feats and movements of the bluefin tuna, *Thunnus thynnus*, were included. This fish is distributed widely in both the Atlantic and Pacific Oceans (Figure 2.9). Although these tuna may be found in all temperate and tropical

Figure 2.9: The geographical distribution of the bluefin tuna (oblique lines). The solid black represents two spawning areas for the Atlantic tuna.

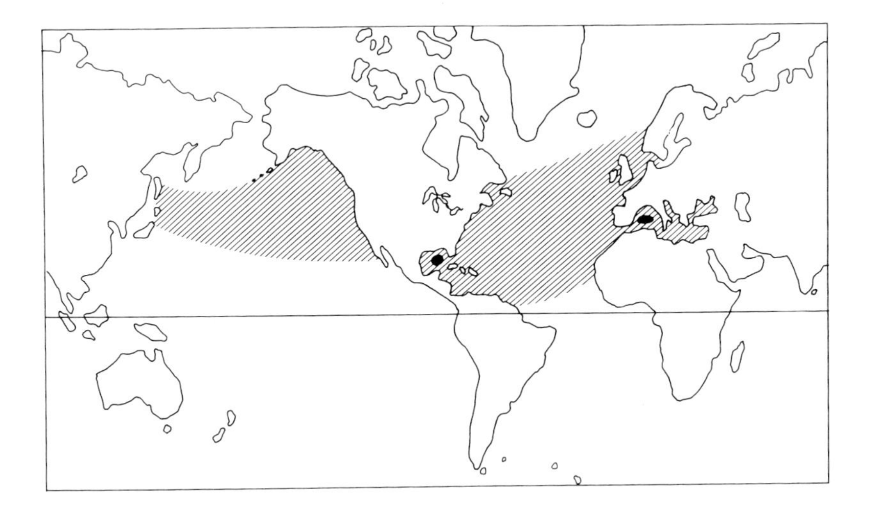

Source: drawn from descriptions presented by Hart, 1973 and Butler, 1982

marine waters of the world, southern hemisphere populations have been regarded as a separate species (Godsil and Holmberg, 1950). In the Pacific Ocean the bluefin tuna ranges in the east from Guadalupe Island off the northwest coast of Mexico north to the Gulf of Alaska (rare north of 35°N latitude) and in the west from Hokkaido and north to the Kuril Island (Hart, 1973). In the Atlantic Ocean the bluefin ranges from the Caribbean Sea north to Newfoundland in the west and from the Canary Islands through the Mediterranean Sea north to the North Sea in the east. The Atlantic Ocean has major tuna spawning areas such as in the middle to eastern Gulf of Mexico, in the Mediterranean Sea from Spain to Italy, outside the Strait of Gibraltar, in the Black Sea and around Bermuda. The different races exhibit a small and variable interchange of individual fish. A few dozen fish tagged in the west Atlantic have, however, been recaptured in the eastern Atlantic, but only half a dozen fish have been reported making the reverse transatlantic migration (Butler, 1982). From these tagging experiments it is interesting to note the distance and speed of bluefin tuna migrations. Seven large fish have been tagged in the Straits of Florida and recovered near Norway, a distance of over 10,000 km. One tagged tuna that accomplished a transatlantic migration with the current averaged over 5 km/hr.

Female blufin tuna carry from 1 million to 30 million eggs depending on their size. The eggs and larvae are pelagic and drift a few meters below the surface. As the fish grow they feed on small pelagic fish such as herring, sardines (*Sardinia pilchardus*), anchovy (*Engraulis encrasicholus*) and mackerel (*Scomber scombrus*). Larger tuna also feed on cod, eels (*Anguilla anguilla*) and larger invertebrates such as cephalopods. Bluefin tuna grow rapidly, reaching 4 kg in their first year and reaching sexual maturity from 3-5 years. An average sized adult of 14 years might weigh 290 kg and be 2.5 m long, although much more formidable fish up to 680 kg and aged 32 years have been recorded.

The migration pattern of the bluefin tuna race that spawns in the western Mediterranean in May and June is depicted in Figure 2.10. Following spawning, some spent fish begin a migration circuit that is completely contained within the Mediterranean Sea. The majority of spawned fish migrate through the Strait of Gibraltar and north as far as Norway. These migrations may be following the surface currents that flow north along the west coast of Ireland and past northern Scotland. In the winter they move south and are found in deeper water, possibly migrating in a counter-current system. In the spring they return to the surface and may employ small gyral currents for feeding before entering the Mediterranean to spawn.

The albacore tuna, *Thunnus alalunga*, is smaller than the bluefin tuna but is also known to follow long migratory paths. This species is distributed widely over the Pacific, Atlantic and Indian Oceans (Figure 2.11). The albacore spawns in tropical waters in spring but migrates north in the later spring and summer. The group of albacore in the Pacific display three separate migration circuits (Figure 2.11). The eastern circuit is comprised mainly of young albacore up to age three years. During the following two or three years the fish change and follow the migration circuit in the west. When the albacore reach maturity at about six years of age, they then migrate south to the spawning grounds. In the Atlantic ocean the spawned adults and one year olds migrate north in February at relatively shallow depths. In the fall, these tuna migrate south from Iceland to the spawning grounds in tropical waters but at depths of up to 350 m.

Littoral Migrations

Littoral migrations take place in fresh water, sea water or between the two, and could be included within the categories of migration referred to earlier.

Figure 2.10: Annual migration circuit of the bluefin tuna stock that spawns in the Mediterranean Sea.

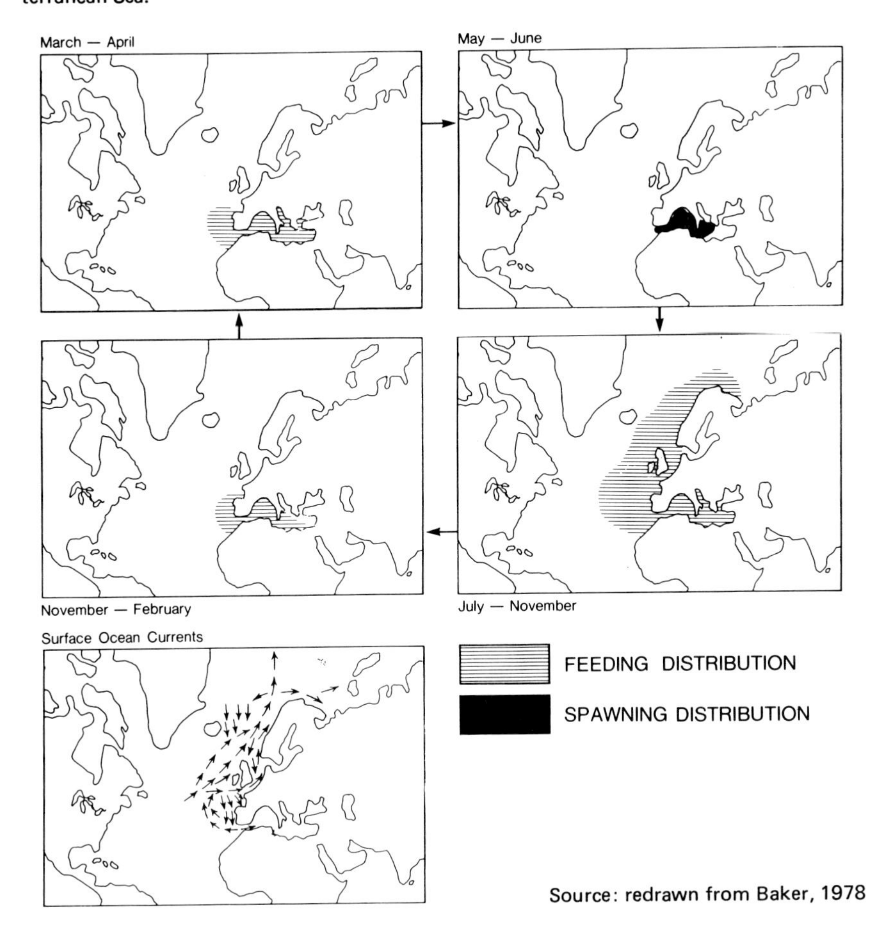

Source: redrawn from Baker, 1978

However, these are somewhat unique types of migration patterns in that they usually cover only short distances and occur between the inshore and offshore and/or shallow to deeper waters. They are usually regular and exhibit daily, monthly (lunar) or seasonal periodicity. Baker (1978) distinguishes these types of migrations and outlines three examples, but he emphasizes that these shorter migrations are more or less species-specific. The migrations of the same three species will be considered here as they are actually short oceanodromous migrations.

The littoral zone that borders the sea and the land provides an array of physical and biological parameters that have given rise to selective pressures for organisms causing them to distribute themselves in a vertical zonation pattern. The factors that contribute to this selection are such things as varying times of exposure to air and water, temperature, light intensity, salinity and

Figure 2.11: The annual migration circuits for two groups of albacore tuna in the North Pacific and North Atlantic Oceans. The North Pacific group displays three annual migration circuits. The eastern circuit is performed mainly by tuna up to age three years. The fish then enter the western circuit and when reaching maturity at six years of age they then migrate south to breed in tropical waters. In the North Atlantic, the post-spawned tuna and yearlings migrate north in February and in the following autumn migrate south. Numbers represent the months when the tuna are in that part of their migration circuit.

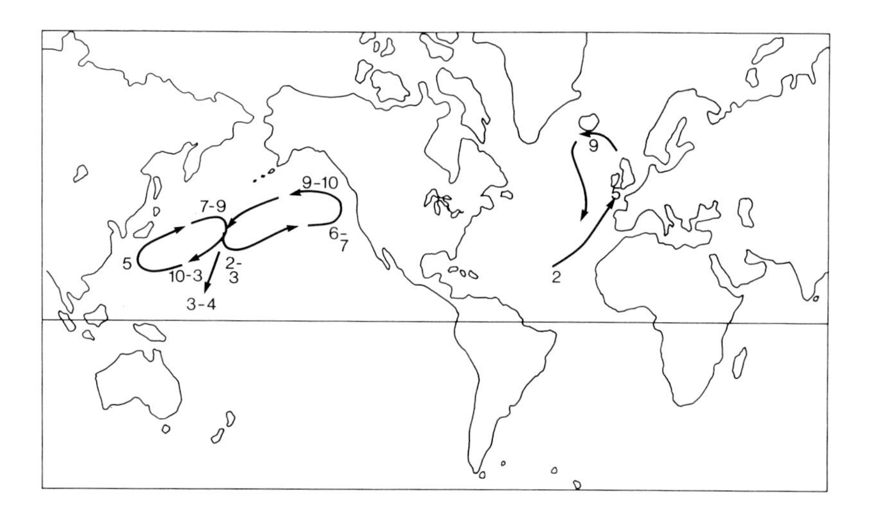

Source: redrawn from Baker, 1978

size as well as the nature of deposited organic particles. Since the tide regularly covers this environment, exploitation of the intertidal zone requires the development of a regular migration pattern that may have to be corrected to the day/night cycle as well as to the ebbing and flooding of the tide. The migration cost, especially on steep rocky shores, must be offset by benefits of food availability, predator escapement or some other attribute of this environment such as desirable spawning beds. Migrants in the littoral zone may move in any one of three different patterns (Baker, 1978). They may maintain a constant position relative to the edge of the water and thus will migrate back and forth as the tide changes. Other migrants may burrow into the sand or hide under rocks or vegetation well up in the intertidal zone and move down the shore to feed. Examples of this type of migration pattern come mainly from the invertebrates. Thirdly, some fish migrate to the intertidal zone only when the tide is high and thus may be considered as sub-littoral animals.

Daily and/or tidal return migrations occur in a number of fish such as different species of blennies, killifish and sculpins. The migrations and physiological adaptations of the Mozambique mud-skipper, *Periophthalmus sobrinus*, are particularly interesting. This fish inhabits areas along tide channels that drain mangrove areas and flow across open mud flats. At low tide the mud-skipper is found in the middle of the intertidal zone foraging in the open but close to a tide channel. It obtains its prey by visual perception and thus feeds during daylight hours. The mud-skipper spends nights in an underground nest. In the winter the fish stays in its nest when the tide is high but in the summer it follows the tide in along the tide channels. When the tide recedes or darkness approaches, the mud-skipper seeks out its underground nest.

Tidal and/or daily migrations occur because of changing environmental parameters that are largely physical in nature. Seasonal littoral migrations on the other hand are usually species-specific and occur due to intra- and interspecific relationships. Seasonal migrations take place to reduce the risk of predation or enhance the food availability, each of which often must be compromised by the other. Gibson (1967) describes the distribution of the blennioid shanny, *Blennius pholis*, on the west coast of Wales during the summer and winter. As the temperature drops in the winter the shanny moves to deeper water where *Laminaria* seaweed is found, in order to seek food. In the summer these fish are found in pools well up in the intertidal zone where the seaweed *Fucus* is found. Migration of the shanny varies in different locations and seems to depend on the different seasonal temperatures. Even within one area the migrations will vary from year to year but appear to be correlated to the different annual temperature regimes. Temperature thus appears to be the ultimate factor changing the migration pattern but the proximate factor may be the changing availability of food.

The southern races of the lumpsucker, *Cyclopterus lumpus*, and the viviparous blenny, *Zoarces viviparus*, migrate to inshore areas of Britain during the winter and move offshore in the summer. These fish move to inshore waters in order to breed and are likely to do so in the winter to reduce the risk of predation. The inshore temperatures are lower in the winter and cause a number of potential predators to move offshore. Although the food availability is lower inshore during the winter, the absence of predators during the reproductive season is more important. California grunion also migrate well up on sandy shores at night during the spring and summer high tides in order to spawn. The eggs develop in the moist sand out of reach of the normal tide. The eggs are washed out by the next extra high tide which coincides with the lunar cycle. Once immersed in sea water the eggs hatch and the young live in pelagic existence until returning to spawn after one year. These migrations were discussed earlier in this chapter and are also covered in more detail in Chapter 3 with respect to their orientation mechanisms. The grunion probably perform these migrations in order to provide the eggs with an environment free from marine predators. Another Atherinid, *Hubbsiella sardina*, performs similar migrations to deposit eggs but spawns during the day instead (Baker, 1978). This migration pattern is surprising since predation of the adults would seem to be high during daylight hours but possibly reproductive success in terms of visual contacts outweighs the risks. Other fish such as certain Osmerids (smelts) like the surf smelt, *Hypomesus pretiosus*, and capelin, *Mallotus villosus*, also perform spawning migrations well up onto the beach for the same strategy of reducing predation risk.

Potamodromous Migrations

The most common types of migration pattern exhibited by fish moving entirely within fresh water are seasonal return movements to spawning areas. This usually involves an upstream migration of spawning adults and a subsequent return downstream to feeding areas. A few species, or races of predominantly upstream-spawning adults, move downstream to reproduce. Other biological as well as physical factors often play a role in determining the direction, distance and timing of freshwater fish migrations.

There is a great diversity in the types of fish that migrate entirely within fresh water. Fish displaying these migration patterns include all three classes of fish-like chordates. Many Cyclostomes belonging to the lamprey family (Petromyzontidae) inhabit only fresh water and exhibit typical migrations from deeper water into shallower spawning areas. Even within the predominantly marine cartilaginous fishes there exists a stenohaline group of freshwater sting rays (Potamotrygonidae) that migrate within the Amazon, the Mekong and Niger rivers. Within the bony fishes there are many groups displaying migration patterns solely within fresh water and these include species of the more primitive sturgeons and paddlefish (Chondrostei), garpikes (Holostei) and many families of Teleostei such as certain shad (Clupeidae), suckers (Catastomidae), minnows (Cyprinidae), catfish (Siluroidei), darters (Etheostominae), salmon, trout, white-fish, grayling (Salmonidae), smelt (Osmeridae), pikes (Esocidae), killifish (Cyprinodontidae), sculpins (Cottidae), cods (Gadidae), perches (Percidae), sunfishes (Centrarchidae), and sticklebacks (Gasterosteidae), to mention only a few. Many of these groups of fish contain species that are entirely stenohaline freshwater forms, whereas others have the ability to withstand salt water. Some species have different races that inhabit only fresh water and yet other races that are anadromous (for example, lampreys, trout, salmon, sticklebacks). The reasons for these differences within a given species are difficult to appreciate especially when the two races are from the same drainage system or even from within the same lake. It is also hard to understand why certain anadromous species when transplanted become completely confined to fresh water even though they have access to the ocean. Such examples exist for the transplants of Pacific salmon species to the Great Lakes system in eastern Canada and the introduction of Atlantic salmon, *Salmo salar*, into Lake Te Anau, New Zealand.

Many potamodromous movements involve spawning migrations of adults upstream and subsequent downstream movements of the spent adults which are followed by similar displacements of the larvae or young. The spawning grounds may be sought out for a number of possible different reasons. These specific sites might not be adequate for supporting the adults but are necessary for the developing eggs and young since they might provide the correct substrate for deposition and future development of eggs. The characteristics of the water flow in these areas may also be important for providing appropriate temperature regimes or oxygen concentrations. The shallower upstream nursery areas may be chosen due to a lack of predators on the eggs and young. The larger fish are possibly excluded due to limited food availability or because of a lack of cover afforded by the shallow water. Often the freshwater spawning grounds might be required because of a higher production of food suitable for the young. The migration of the adults away from the spawning areas after reproduction may be brought about by changing environmental conditions that make the headwaters unsuitable during different parts of the year. In north temperate zones shallow upstream areas are the first to be affected by lowered temperatures in the winter making this environment unsuitable during freeze up. In the summer months many headwater areas dry up or have much reduced water flow, again making such habitats unsuitable for large adult fish. Upstream spawning migrations may occur due to the pressures of migration cost. Larger fish can accomplish upstream migrations at a lower energetic cost relative to their size than can smaller fish. Thus, the movement of young fish to feeding areas is likely to be more successful if the fish can move downstream with the current. From an energetics point of view it is also likely that migration patterns of fish in fresh water are return migrations to natal areas. Upstream movements are energetically costly and if fish had to search for appropriate spawning sites the migration cost could be too great. Spawning success is thus more likely for a migration pattern to a predetermined site where the spawning ground was used successfully by preceding generations.

Migrations of fish within fresh water may also occur for reasons other than spawning. In the freshwater environment fish may be forced to move to other areas due to a number of factors that are largely biotic in nature. Population density may become too great and thus result in intraspecific competition. Less fit individuals may, therefore, be forced to emigrate to less desirable areas. During intraspecific contests, individual fish may be displaced by currents – a form of passive migration. Food availability may also change seasonally and force large numbers of fish to seek out new, more abundant food sources available in other areas. The availability of space as shelter from the current or from predators imparts pressures on individuals to migrate to more suitable environmental territories.

Although there are many species of fish that migrate solely within fresh water, as mentioned earlier, their migration patterns are similar and take place for much the same reasons. To give examples of potamodromous migrants therefore, would only be a recapitulation of the general pattern already discussed. Migration patterns of specific anadromous fish will be discussed later. These migrations might be thought of as extensions of movements to feeding areas that happen to be in sea water. The freshwater phase of these migrations is similar to that of stenohaline freshwater fish and these can therefore be referred to as examples of freshwater migration patterns. In addition, the four groups of fish chosen to represent anadromous migrants also contain races that are entirely restricted to fresh water and exhibit migration patterns within that environment only. Spawning migrations within fresh water are usually upstream but occasionally downstream, or if within a lake occur generally from deeper to shallower water. Displacement of young or subordinate adults to feeding areas usually occurs downstream but some examples may be found that are in the reverse direction. Migrations within fresh water may appear to be short displacements to feeding areas or spawning sites but some species exhibit long-distance migrations that are formidable physiological feats. The omul, *Goregonus autumnalis*, migrates 1600 km up the Lena River, USSR and the sturgeon, *Acipenser sturio*, is known to move up the Danube to spawn a distance of 2000 km (Hynes, 1970).

Diadromous Migrations

Some oceanodromous and potamodromous species have evolved into catadromous and anadromous species respectively. Fish that migrate solely within the marine environment generally complete migration circuits and normally take advantage of ocean currents to reduce migration costs as well as possibly to use an orientation cue. Fish such as eels that are catadromous and enter fresh water to grow before returning to the sea to spawn, display migration circuits similar to oceanodromous species. Fish that migrate entirely within fresh water generally move upstream or into shallower lake waters in order to spawn; this is followed by a downstream movement of young where growth and development into adulthood will take place in feeding areas. Anadromous species that have entered the sea water environment to feed still return to fresh water to spawn and the movement patterns within fresh water are very similar to a potamodromous species.

When a fish migrates within either fresh water or sea water it is not difficult to imagine that when the freshwater/ocean barrier is reached from either direction, some species have found it advantageous to cross this barrier and continue their growth in a different salinity environment. However, as Baker (1978) points out, it is difficult to appreciate why some species moved from a predominantly freshwater environment into ocean feeding areas whereas others have moved in the opposite direction from a marine environment to feeding areas within fresh water. It is particularly difficult to understand how catadromous and anadromous behavior developed in the same drainage system. Migrant fish may have become diadromous due to both inter- and intraspecific selective pressures. Fish in both the freshwater and ocean environments have evolved along with various parasites and predators. If a given species could evolve into a fish with euryhaline abilities faster than its parasites or predators, it could then enter a new salinity environment and thereby escape species-specific diseases and predation. Such selective pressures could impinge on freshwater as well as marine species. Inter- and intraspecific competition for various resources in the freshwater or sea water environments could also act in a similar manner to select for diadromous behavior.

Baker (1978) has summarized the world distribution of fish that display diadromous migrations. Although there is an overlap, the majority of anadromous migrations occur in cold-temperature and sub-polar regions, whereas catadromous migrations are more evident in tropical and warm-temperature regions. This latitudinal or climatic effect on diadromous migrations is not likely to be due to effects within fresh water or the ocean. The general pattern of upstream-downstream migrations of potamodromous species appears to be similar irrespective of latitude. Oceanodromous migratory circuits likewise appear to be independent of latitudinal effects. The critical point for diadromous migrations that imparts a latitudinal effect is likely to be at the boundary between fresh water and the sea. Thus, selection in the colder, higher latitudes appears to favour movements of some fish from the freshwater environment to the ocean, whereas in warmer, lower latitudes, there seems to be a bias towards the selection of some young fish to move across the salinity boundary into fresh water. The actual factor or factors that cause an advantage for growth and development at low temperatures in the ocean or higher temperatures in fresh water, are only speculative at this time. Temperature is the likely environmental parameter that would cause the latitudinal variation between anadromous and catadromous behavior. Temperature may be the proximate factor, but may only be a reflection of the ultimate cause due to its effect on some energetic or physiological process such as efficiency of osmoregulation or food availability.

Diadromous species exhibit varying degrees of return to their natal areas. Fish species that have a low degree of return may thus lose their diadromous behavior: catadromous species might therefore reside entirely within fresh water and anadromous species might become completely marine for their entire life cycle. There are many families of present-day fishes that contain both freshwater and marine species. The ancestors of these groups of fish presumably evolved in either a freshwater or marine environment. Over time some species found it advantageous to cross the freshwater/seawater barrier and thus become diadromous. Some of these migratory fish in turn experienced selective pressures to remain in the new salinity environment throughout their life. Species may have thus changed many times from being stenohaline marine to euryhaline to stenohaline freshwater to euryhaline and so on. Tchernavin (1939) describes salmonids as probably being originally marine species when they were more closely related to Clupeids (herring, pilchard, shad). Later they likely became catadromous and subsequently only freshwater. Many salmonids today are freshwater species but many are also now anadromous. Some of the salmonids such as pink salmon, *Oncorhynchus gorbuscha*, which are almost entirely anadromous, have some races that only migrate into estuarine areas to spawn. The entire species in fact actually has a shorter freshwater phase in early life stages than most other Oncorhynchids. This species may thus be in the process of becoming oceanodromous again.

Catadromous Migrations

Fish that spend most of their lives in fresh water but return to the ocean are few in numbers compared to anadromous species and their migration patterns are less well documented than for other migrants. There is some information on migrations of jollytails or trout of the southern hemisphere (*Galaxias* spp.) but most data on catadromous migration patterns come from studies on eels belonging to the genus *Anguilla*.

Eels. The freshwater eels (Anguillidae) are represented by about 16 different species of the genus *Anguilla* that are distributed widely throughout the world (Figure 2.12). They inhabit drainage basins flowing into the western and eastern North Atlantic Ocean extending to Norway and north-western USSR. They are also found in various locations from streams around the perimeter of the Indian Ocean, in southeast Asia and in the western South Pacific including New Zealand and Australia. In the North Pacific, *A. japonica* is thought to spawn south of Japan and in the South Pacific *A. australis* and *A. dieffenbachi* seem to spawn east of northern Australia. The western North Atlantic (Sargasso Sea) may be

Figure 2.12: Three possible migration circuits of freshwater eels. The black areas represent spawning areas and the horizontal lines indicate freshwater feeding areas. Arrows represent surface ocean currents. The North Pacific area is frequented by the Japanese eel, *Anguilla japonica*. The South Pacific area is used by *A. australis* and *A. dieffenbachi*. The former species occurs in larger lakes and rivers than does the latter. The North Atlantic spawning area may be used by the European eel, *A. anguilla* as well as the North American eel, *A. rostrata*.

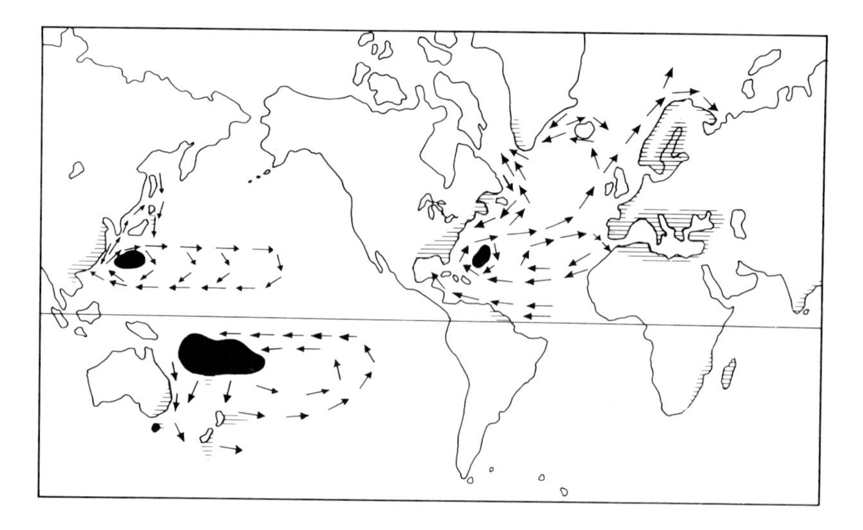

Source: redrawn from Baker, 1978

the spawning grounds of either *A. rostrata* or *A. anguilla* or both but this is still a subject of ongoing controversy. This area has not been confirmed by the presence of eggs, small larvae or spawning adults.

Freshwater eels are thought to spawn at ocean depths of 400-700 m but in water of much greater depths. The eggs are pelagic and, unlike many fish species, development then proceeds through a distinctive larval stage known as the leptocephalus, which has a transparent, ribbon-like body and conspicuous eyes. The larvae drift with ocean currents for periods of one to three years depending on the species and eventually move towards the estuaries of rivers. As the larvae approach inshore waters they metamorphose into small, transparent fish called glass eels. Subsequently, pigmentation develops by the time the eels enter fresh water when they are at a size of 50-100 mm. The young eels or elvers, as they are called, migrate upstream and eventually lie buried in muddy or silty benthic environments during daylight hours. The eels feed mainly during the night on a diet consisting of small fish, insect larvae, crayfish, snails and earthworms. Eels take several years (one is recorded having an age of 88 years in captivity) to mature in fresh water before initiating a spawning migration. Maturing adults are thought possibly to return to natal areas where they spawn but little information is available for many species. The adult migrations may complete oceanic migration circuits by choosing appropriate gyral or countercurrents but more information is required before this can be well established.

Schmidt (1922) used special nets to capture small eel larvae. By comparing the length of larvae, place of capture, time of capture and currents, he estimated that they originated from eggs deposited southwest of Bermuda in an area often called the Sargasso Sea. Although the larvae drift passively in the horizontal ocean currents they do undertake daily vertical migrations. Baker (1978) compares the distribution of larvae in the North Atlantic with ocean current patterns emanating from the Sargasso Sea (Figure 2.13). Currents from the southern portion of the suspected spawning area would cause larvae to be carried to North American offshore areas. Currents from more northern areas of the Sargasso Sea flow directly to northern Europe but would take two years to carry the larvae such a distance, whereas the larvae would arrive at North America in only one year. Other ocean currents from the northern section of the Sargasso Sea would take larvae on a circuitous route and delay subsequent arrivals in Europe by another year or carry larvae to other eastern Atlantic locations. It is thus possible that American eels are derived from spawnings in the southern Sargasso Sea, whereas European eels are from the northern portion. Since the American eels arrive at river mouths one or two years earlier than European eels, their metamorphosis to elvers must occur earlier. Growth rates and morphological characteristics appear to be distinct in these two groups. The time of arrival of metamorphosed eels appears to coincide with estimated times larvae would require to drift to the different freshwater sources. If indeed larvae arrive and metamorphose at different times, mechanisms must be present to cause the transformation into elvers and allow the migration to continue into fresh water.

Figure 2.13: The possible spawning area of the North Atlantic eels (square box). The southern portion of this area has surface water flowing out in a direction indicated by the dashed line whereas water from the northern portion moves in the direction represented by the solid line. M, J, S, and D indicate March, June, September and December respectively. The suffixed numbers represent the years after spawning. The dotted lines indicate times of arrival of eels at various points off the European coast. The numbers by these lines indicate the month of arrival.

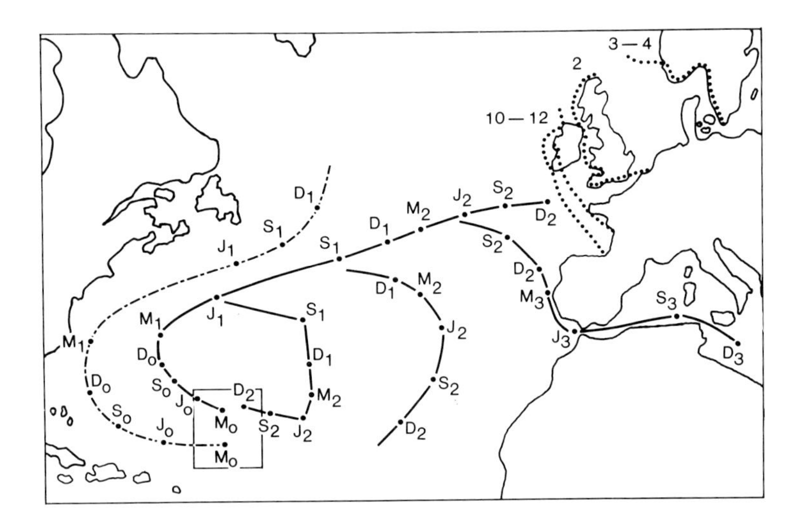

Source: redrawn from Baker, 1978

If larvae metamorphose as they approach the continental shelf or river mouths, there may be some stimulating factor in these waters that enhances transformation or alternatively, there may be some inhibitory factor in the marine environment. If the various geographical areas of freshwater distribution represent different species or races of eels that are also reproductively isolated, there could be a genetic basis to the time at which metamorphosis occurs. Serological studies by de Ligny and Pantelouris (1973) on American and European eels suggest that there are genetic differences between these groups.

There is still uncertainty about the existence of more than one *Anguilla* species in the North Atlantic Ocean. Once adults become silvered and enter the ocean their migration routes to spawning grounds are not unequivocally known. Schmidt (1922) suggests there are two distinct species, one spawning in the north and the other in the south of the Sargasso sea. On embyrological, morphological and serological grounds, this seems to be a viable hypothesis. On the other hand, Tucker (1959) proposes that adult European eels die before reaching the spawning grounds and the entire following generations in both America and Europe are derived from spawning American eel populations. The subsequent morphometric differences between populations may only be a reflection of different environmental parameters experienced by each population.

Alternatively, the spawning populations could possibly be composed entirely of adults from Europe. These last two possibilities appear to be energetically unsound from an evolutionary point of view but definitive data in the form of the presence of adults in the spawning grounds are still elusive.

Anadromous Migrations

Many different families of fish exhibit migrations from freshwater natal areas to feeding areas in the ocean followed by a return spawning migration back to fresh water. Although this general migration pattern occurs in a large number of different species, the degree of return to natal areas, the distance travelled, the timing of various phases of the migration and the obstacles confronted vary considerably between species and even within certain species.

Salmonidae. This family ranges widely throughout the northern hemisphere including well north into the Arctic. A number of species have been successfully introduced into cool southern hemisphere environments. The three subfamilies Salmoninae (trout, salmon, chars), Coregoninae (whitefish, ciscos) and Thymallinae (graylings) contain species exhibiting varying degrees of anadromous behavior. Many of the anadromous species have populations that migrate only short distances into the sea, and others that migrate great distances into the ocean for extended periods of time, or yet others that never leave fresh water.

The Salmoninae is the most commercially important subfamily and the three genera *Salmo* (trout, including Atlantic salmon), *Oncorhynchus* (Pacific salmon) and *Salvelinus* (char and lake trout) have been extensively studied and display varying degrees of anadromous behavior. Many of these migrants display extensive ocean migrations to feeding areas. The sockeye salmon moves from the western shores of North America across the North Pacific almost to the shores of the Kamchatka Penninsula, USSR, whereas others such as the coho salmon, *Oncorhynchus kisutch*, migrate much shorter distances out into the marine feeding areas. The Pacific salmon remain in sea water for varying periods of time depending on the species or population, after which they ascend rivers to their natal spawning areas, reproduce and die. The trout, such as steelhead, *Salmo gairdneri*, and Atlantic salmon can also migrate considerable distances to marine feeding areas, although many others such as brown trout, *Salmo trutta*, and cutthroat trout, *Salmo clarki*, migrate only short distances into coastal feeding areas. The genus *Salmo* also differs in that it is capable of making more than one migration circuit and repeat spawn a number of times. *Salmo* and *Salvelinus* species show a rich variety of migratory patterns from the anadromous to the potamodromous, whereas *Oncorhynchus* species are almost entirely anadromous. Notable exceptions are the landlocked sockeye salmon or kokanee and other Pacific salmon that have been introduced into new environments such as the Great Lakes in North America. The life cycle of the kokanee is similar to that of sockeye except that they remain in fresh water for their entire life cycle. The kokanee are termed landlocked but are so only by choice since they usually have free access to the ocean. It is interesting to note the differences in size between the kokanee and sockeye at sexual maturity. In spite of much greater energy expenditure during the prolonged migration at sea and in river, sockeye are much larger than kokanees. From this comparison, the advantages of anadromous behavior providing enhanced food supplies, becomes quite evident.

The salmonids evolved in cold northern freshwater habitats. Subsequently, anadromous behavior may have developed due to the nutrient-poor freshwater habitat as well as the frequent disappearance of the native fresh water due to glaciation and other factors. The development of specialized behavior and physiological mechanisms necessary for anadromous migrations and the ability to home to natal streams has led to the rapid establishment of reproductively isolated populations. The coho and sockeye salmon have a transformation stage of the fry into a smolt and a much longer freshwater residency. An extreme development of the sea water phase is seen in some species of *Oncorhynchus* which spend only a small proportion of their life in fresh water. Pink, *O. gorbuscha*, and chum, *O. keta*, salmon generally spawn in sections of rivers close to the ocean and produce young fry that migrate to the sea shortly after hatching. The pink and chum salmon also display stronger nocturnal schooling behavior with little hiding and downstream orientation. Individual species within the genus *Salmo* such as *S. gairdneri*, also display tremendous variability with respect to anadromous or potamodromous behavior. Genetic as well as environmental factors seem to be important in determining direction and type of migration pattern followed (see Chapter 3).

Sticklebacks. The three-spined stickleback, *Gasterosteus aculeatus*, has a wide distribution including both shores of the North Pacific as well as the eastern and western North Atlantic. It is found in freshwater and marine environments and is occasionally recorded in larger numbers far off to sea. These sticklebacks are found as far south as the Tropic of Cancer and north to the Bering Sea, Greenland and Norway. In the northern parts of its distribution the three-spined stickleback is almost entirely marine, only entering fresh water to breed. In the south it is almost entirely confined to fresh water. The development of anadromous behavior may thus be due to the lower temperatures in the higher latitudes.

The anadromous form (*trachurus*) is morphologically different from the freshwater form (*leiurus*). The anadromous form only enters fresh water in the spring to breed and has structural modifications reflecting its migratory behavior. *Trachurus* is thus more slender and streamlined, possesses a greater number of long and slender gill rackers for planktonic feeding, builds its nest in currents and produces more eggs. The anadromous behavior of sticklebacks may be an advantage in the sense that they can exploit a larger food resource. The larger size of *trachurus* would tend to support this idea. The anadromous form also has many more lateral plates which may be an indication of different predation pressures. Their relatively small size, availability and ease of handling, makes three-spine sticklebacks an attractive species for investigations into the development of anadromous behavior as well as for physiological studies of migration with respect to osmoregulation and mobilization of energy reserves.

Shad. The American shad, *Alosa sapidissima*, is found in the western Atlantic and was introduced to the Sacramento River, California, USA in the early 1870s. As with other Clupeids such as herring, the shad undertakes extensive oceanic migrations. The populations introduced into California produced future generations in Alaska only 30 years later. The rapid spread of this species in the Pacific is good evidence of substantial ocean movements. Tagging experiments in the Atlantic indicate extensive and rapid movements as well.

Shad seem to be able to home to natal streams by the use of olfactory and rheotropic mechanisms (Dodson and Leggett, 1974). A small proportion of adults (10 per cent) may spawn more than once. The young leave fresh water shortly after hatching so that most of the life cycle is in the marine environment. Shad appear to migrate into fresh water depending on the temperature (Leggett and Whitney, 1972; Leggett, 1976). They thus migrate earlier in the year (January) in southern latitudes but not until June in more northerly locations. The correlation between temperature and movement into fresh water may be a reflection of the energetic cost of migration. Shad do not feed in fresh water and consequently lose up to half their body reserves on spawning runs. This may be a reason for such high post-spawning mortalities. Shad seem to be evolving towards a similar life cycle as that of the Pacific salmon.

Lampreys. Hardisty and Potter (1971) have described the life cycles and biology of a number of lamprey species. Some are anadromous and spawn in streams at specific times of the year such as the sea lamprey, *Petromyzon marinus*, while others such as the European lamprey, *Lampetra fluviatilis* are less synchronous. The energetic costs of migration are high and adults die after spawning. The young stay in the streams for a number of years as ammocoete larvae where they lie buried in the mud and are filter feeders. They then metamorphose and during transformation do not feed. If the species migrates to the sea the lamprey become parasitic and feed on fish before returning to spawn about two years later. Some species transform into adults that are nonparasitic. These nonparasitic species have a longer larval life and once metamorphosed do not feed but instead move upstream to spawn. The parasitic species may be able to pursue such a life cycle due to their movements into large lakes or anadromous movements into the ocean where feeding may be possible to attain larger body sizes to enhance not only the energetics of migration, but also of spawning.

3 ORIENTATION

Introduction

Anyone who has witnessed the return of a long distance migrating fish is indeed fortunate. It stimulates the mind to speculate and appreciate the tremendous drive or instinct required to initiate such travels as well as the physical endurance necessary for often long and arduous journeys. For many species of fish, the points at which a migration starts and ends are very distant from each other and the fish may encounter a variety of obstacles and environmental parameters on the way. Such migrations have stimulated the casual observer as well as the experienced investigator to ask many questions as to why a fish would undertake such migrations, what initiates such movements, how the fish can find their way, how the fish knows where to go and the ways and means by which the fish recognizes when it has reached its goal or home.

This chapter will be concerned with how a fish can find its way during the course of its migration. This involves reviewing not only how the fish is able to orient to various directional cues in its physical, chemical, and biological environment, but also what factors are involved in stimulating the fish to move. In addition data are now available on how a fish can recognize when it has completed its migration and arrived at this assumed home or goal. However, although the migrating fish must probably be able to determine where they want to go, relatively little information is available as to the mechanisms of how a 'map' is stored or how an individual acquires it.

Piloting, Orientation and Navigation

The mechanisms of orientation during fish migration have been recently reviewed by Hasler (1971), Leggett (1977) and Able (1980). As has been noted in previous reviews, there is a need to define precisely the term orientation with respect to the phenomenon of fish migration and other types of movement (Figure 3.1). The original categorizations of Griffin (1955) will be used. *Orientation refers to the settling or arranging of an animal in a given compass direction.* A fish would possess the ability to orient if it could maintain a given compass direction (even in the absence of sensory cues emanating from the home site). This may occur in an area familiar to a fish as well as in an unfamiliar area. A type of movement that may be confused with orientation is when a fish can find and recognize its home site by a direct sensory stimulus such as vision or olfaction. This aspect of fish migration is often referred to as *homing* but is more accurately termed *piloting*. It differs from orientation in that the fish may follow some direct sensory cue emanating from the home site that may cause the fish periodically to change compass direction during its journey from some

Figure 3.1: Navigation, orientation and piloting. As a fish migrates towards some goal such as the mouth of a river as depicted in A, it may use one of the three mechanisms. If the fish is displaced to B it may continue on in the same compass direction as before (broken arrows) which would indicate that it was migrating using an *orientation* mechanism. However, if the direction of travel changes towards the mouth of the river (solid arrows) the fish may be *navigating* since it appears to be identifying a specific area. The fish might also be *piloting* by simply following the odor emanating from the river (stippled area). If the fish is displaced to C then the difference between navigation and piloting becomes clearer. At C the broken arrow indicates orientation along the same compass direction as in B. The change of direction towards the river-mouth would indicate navigation mechanisms since the fish is now outside the area of river water odor detection. If the fish was migrating by piloting along an odor trail it would be 'lost' or start a random search(?) for the odor in order to continue piloting.

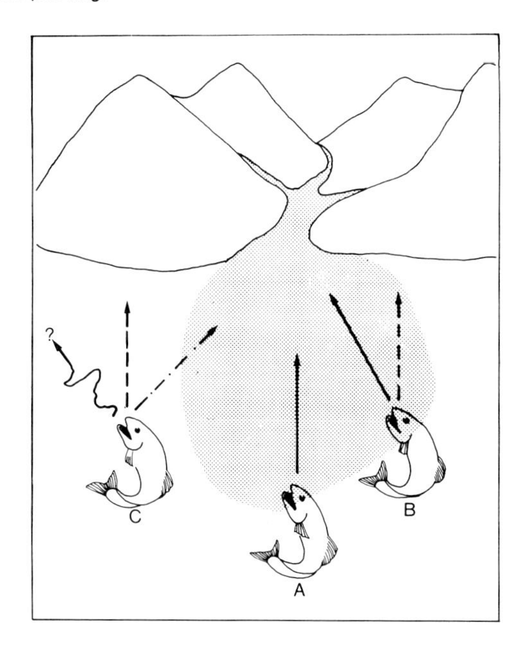

distant point to the home site – even though it is using the same cue throughout this part of its migration. One aspect of piloting is the random nature of the search for familiar cues. In addition, piloting would also lead to a change in compass direction if the fish were displaced. As long as the sensory contact with the home site was maintained, the compass direction would change accordingly so the fish would arrive home following the trail of some sensory cue. If the fish were displaced far enough so that direct sensory stimulation was not available, it would not be possible for the fish to change compass direction towards the home site through this recognizable sensory stimulus.

Navigation is another term used in animal migration, but it differs from both orientation and piloting. Navigation is a mechanism whereby an animal can find the direction towards a given area such as a home site. Unlike orientation the animal has the ability to change its compass direction towards this point in space when displaced. Navigation also differs from piloting in that a change of direction towards this point in space or goal is still possible even if the animal is displaced to some point that is unfamiliar. However, once a new direction is established during navigation following a displacement to an unfamiliar area, the animal may again continue along a given compass bearing by means of orientation until a new position fix is taken. True navigation may be possible in certain species of fish but there is no convincing experimental evidence as yet to suggest any mechanism. To prove whether or not fish possess a navigational ability, experiments that involve displacement of the fish to an unfamiliar area need to be conducted or a simulated displacement by changing the appropriate environmental cues. Unfortunately, the few experiments involving displacement of fish that have been conducted were designed such that the fish still had sensory contact with the home site or other areas with which the fish were familiar; thus the observed homing behavior could be ascribed to piloting as well as to navigation.

Early studies that followed white bass, *Roccus chrysops*, after displacement to open water from the spawning grounds in Lake Mendota (USA) by observations of floats attached to the fish, indicated that when the fish was close to the capture site it probably used olfactory cues for recognition. In open water these fish seemed to have difficulty finding the spawning grounds on overcast days or if they were blinded by opaque plastic eye-caps. This may thus add support to the hypothesis of the use of solar orientation cues (see subsequent sections). Since the whole of Lake Mendota was probably familiar to these fish, navigational ability may not be required (see Hasler, 1971). Other studies employing displacements within familiar ranges of the longear sunfish, *Lepomis megalotis* (Gunning, 1959); European minnow, *Phoxinus laevis* (Hasler, 1966); tidepool sculpin, *Oligocottus maculosus* (Khoo, 1974) and radiated shanny, *Ulvaria subbiturcata* (Goff and Green, 1978) in conjunction with blocking of sensory reception, all indicate that visual and olfactory cues are important in recognizing and possibly orientating (not necessarily navigating) to the home site. Although studies that have employed ultrasonic tracking techniques of displaced adult Pacific salmon have given valuable information concerning individual fish movements and might suggest the use of orientation mechanisms, they do not prove or disprove the presence of navigational mechanisms. This is because the fish probably still have sensory contacts, such as odors or currents emanating from a familiar area traversed recently by adult fish or when the young fish migrated to the ocean as a fry or smolt (Madison *et al.*, 1972; Stasko *et al.*, 1973). There have been a number of experiments conducted in which fish are removed from their natural environment and placed in experimental tanks in order to investigate possible orientation mechanisms. For example, when bluegills, *Lepomis macrochirus* (Goodyear and Bennett, 1979) and sockeye salmon, *Oncorhynchus nerka* (Quinn, 1980; Quinn and Brannon, 1982) were subjected to an artificial shift in phase of the sun and magnetic field respectively, they compensated by an appropriate shift in their mean compass direction. These fish thus appear to be orientating to cues which may be potential factors used for navigation. Many environmental factors both alone and in combination are now known to be used by fish for orientation and these will be discussed in detail in the following pages. It would merit further investigation to determine if one or a combination of many of these factors were responsible for navigational mechanisms in particular fish species.

Quinn (1982b) has recently described a possible model for salmon migration on the high seas. He feels that the established movement patterns of anadromous adults seem best explained by an active orientation system. He proposes a combined map-compass-calendar system to explain the observed movements. A bicoordinate grid of magnetic inclination and declination isolines may constitute the map. Compass mechanisms such as celestial and magnetic orientation are proposed to maintain the headings determined from the map sense. An assessment of daylength or the rate of change in daylength may combine with the endogenous circannual rhythms of the salmon to provide a calendar sense. Quinn presents the ranges of inclination (Figure 3.2), declination (Figure 3.3) and daylength in the North Pacific region and proposes two types of experiments to test his map hypothesis.

Figure 3.1: Inclination isolines of the earth's magnetic field in the North Pacific Ocean.

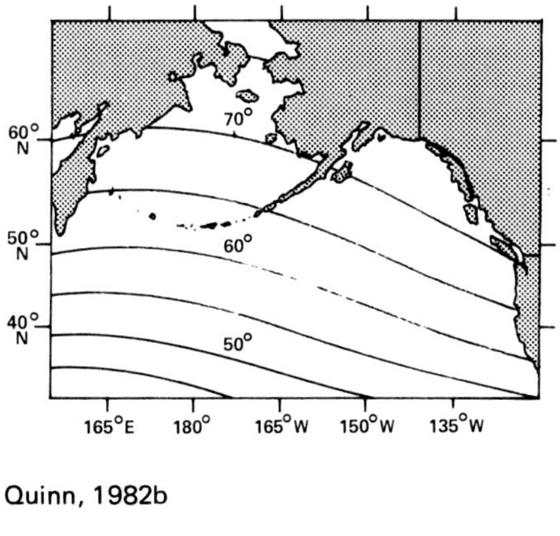

Source: redrawn from Quinn, 1982b

The Construction of a Map

Many species of fish migrate over long distances and can return to specific predetermined sites. During such long distance movements the fish require input from their environment to orient themselves to this goal. Even though these fish may exhibit sophisticated orientation mechanisms, additional information or knowledge to know where to go may be advantageous, i.e. the possession of some kind of 'map'. The information required for this map might be stored in neuronal circuits in the central nervous system and obviously requires appropriate acquisition capabilities. The source of a map might be instinctive (genetically

Figure 3.3: Declination isolines of the earth's magnetic field in the North Pacific Ocean.

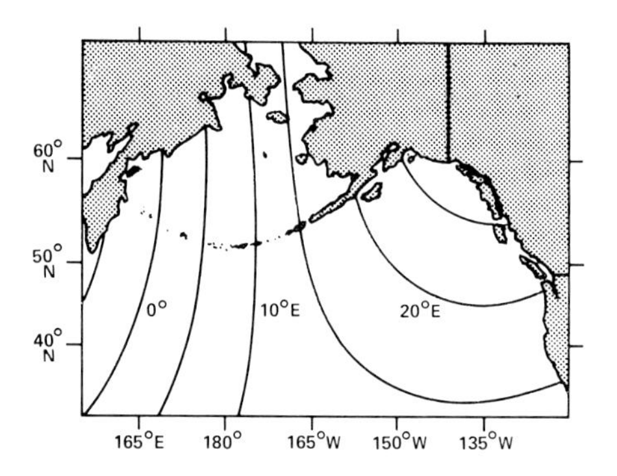

Source: redrawn from Quinn, 1982b

determined), or may be learned during ontogeny and adult stages of the life of the individual. If such a map is learned, it may be acquired by a number of different types of learning, including imprinting. Imprinting is a rapid acquisition of certain characteristics of a stimulus object that need not be reinforced and usually occurs during a short period of time early in the life of an animal. There is, however, considerable controversy over whether imprinting is a special form of learning (perhaps involving innate mechanisms) or is similar to other kinds (see Bateson, 1979; Horn, 1979).

Inherited Mechanisms. By studying fish orientation abilities at early stages of ontogeny some information has been gained with respect to the genetic or learned components of these mechanisms (Brannon, 1972; Quinn, 1980; Brannon *et al.*, 1981). Quinn (1980) tested sockeye salmon fry from two different populations shortly after emergence from gravel and subsequent migration of the fry to their respective nursery lakes where they remain for a year before migrating to the sea. Both groups exhibited orientation preferences in laboratory arenas in the appropriate compass directions that would allow them to migrate in their nursery lakes. He concluded that fish orientation might have been genetically determined or learned from river migration. Subsequent work by Brannon *et al.* (1981) tested sockeye salmon fry that were newly emerged before lakeward migration and once again the fish oriented on a specific compass direction. Since the fry oriented immediately after emergence from the gravel, they could not have learned the correct direction during movement down the natal stream and so the authors concluded that the fry possessed an inherited orientation ability. However, the possibility still exists that fish were exposed to directional cues while still in the gravel that entrained later orientation. Interestingly, this latter study also showed that the orientation could be modified appropriately by odors from nursery lake water as the fish grew older, implying that an inherited response can be triggered by environmental and temporal cues.

More definitive research has been initiated by Bams (1976) with respect to the inheritance of orientation mechanisms. He carried out an experiment with pink salmon, *Oncorhynchus gorbuscha*, that compared the time of migration and numbers of returning adults between a pure donor stock and a hybrid stock created by crossing females from the donor with males of the local residual stock. The fry of the pure donor stock migrated to the sea earlier than the hybrid stock but both groups of adult fish returned at the same time. Thus, timing of specific parts of migration appears to be genetically determined. The number of returns was much higher for the hybrid stocks than pure donor (non-natal) stock. The difference between the two groups was ascertained by the author as not apparently due to differential mortality. The homing ability, possibly orientation, seems to be partially due to inherited factors and not entirely to learning or imprinting.

Brannon (1967, 1972) and Rayleigh (1967, 1971) initially reported that there was a genetic component to migratory behavior of newly emerged sockeye salmon fry. Northcote and co-workers have more recently reported on genetic aspects of rainbow trout, *Salmo gairdneri*, orientation by rheotropic responses (Kelso *et al.*, 1981; Northcote and Kelso, 1981) (Figure 3.4). In order to maintain trout populations above impassable waterfalls, emerging fry need to respond with a positive rheotaxis. Likewise, emerging trout fry from inlet and outlet streams of a lake used later by the adult fish must possess an appropriate and opposite rheotaxis in order to arrive at the lake. These studies on directional

Figure 3.4: Response to water current shown by two homozygous LDH phenotypes of rainbow trout fry in experimental tests permitting upstream or downstream movement Ha^A phenotypes showed significantly more upstream movement than Ha^B phenotypes in tests made under lighted conditions and less downstream movement in darkness.

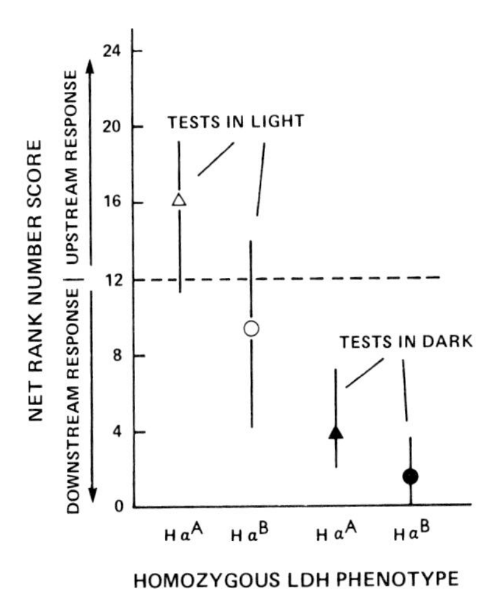

Source: redrawn from Northcote and Kelso, 1981

responses to water currents by fry from parental stocks that spawn either in inlet or outlet streams and either above or below waterfalls, lend considerable support to inheritance of rheotropic responses. It is also interesting to note that these trout from fast flowing waters exhibit lactate dehydrogenase genotypic differences. The genotypes can be differentiated not only by serum electrophoresis but also by endurance tests. The authors discuss an intriguing possibility that there is a genetically determined enzyme system that can alter the physiological capabilities of the fish which might be associated with the current response. These studies also indicate that the current response can be modified by such environmental factors as illumination and temperature. Although these studies demonstrate a genetic contribution to some aspects of migration such as specific orientation abilities, it is not known whether or not a type of map could likewise be inherited.

Learned Maps. The studies described above indicate a genetic basis to orientation. In addition, the mechanism appears to be capable of modification by appropriate environmental stimuli. This orientation may be only an expression of genetic information but it may, in addition, be an example of learning required for constructing a map for migration.

Baker (1978) describes possible movement patterns of fish that might be important for the learning and retaining of a map for later migrations. Early in the life cycle of fish species, various habitats are perceived. The fish can easily move from one point to another in this familiar area since it has prior information as a result of direct perception. Two types of long range movements away from this familiar area have been described by Baker. One of these types he calls 'non-calculated removal migrations', which are movements beyond a familiar area during which information is not retained for a return migration. The other type, termed 'exploratory migration', is of interest with respect to the establishment of a map since spatial and sensory information may be retained throughout so that the fish may have the ability to return to the familiar area. The exploratory migration may, however, be in a linear form and thus might be thought of as only an extension of the familiar area. The information that is learned and stored from exploratory migrations may be in the form of a spatial or a sequential memory and may be gathered by the entire sensory repertoire of the fish. A spatial memory may be an advantage to a fish since the familiar area may change during the hiatus between an exploratory migration and the subsequent return but the sequential memory may be simpler. The use of either may thus give selective advantage to different fish species. There is evidence from investigations on certain salmonids that at least part of the memory is sequential since there is a signficant reduction in the homing success of upstream migrants when they are transported over a portion of their previous downstream migration (Carlin, 1955; Wagner, 1969; Brannon, 1982). Baker's hypothesis would allow all types of fish to have access to a storage of information in a memory that would constitute a map. Not only could the fish employ appropriate environmental cues for this memory to find its way, but it might also allow the fish to be capable of learning routes that minimize migration energy costs from one point to another within the familiar area.

Imprinting. Imprinting could be a mechanism for obtaining part or all of a stored map. The use of olfactory cues by adult fish returning to their spawning grounds is well documented (see Leggett, 1977; Hasler, 1983) and may be an example of at least part of a map present for migration orientation. The detection and use of odors by migrating fish as well as the various types and sources of these odors, will be discussed later. It is interesting to note at this point, however, that fish migrating to their spawning grounds can recognize odors from their natal stream as well as synthetic chemicals that were present in the water at early stages in the life cycle of the fish (Scholz *et al.*, 1976; Hasler *et al.*, 1978; Johnsen and Hasler, 1980). Such information is valuable in designing experiments to indicate how an olfactory cue can serve as a sign stimulus to release an appropriate behavioral response to other orientating cues. The recognition of a particular odor (sign stimulus) may thus elicit the appropriate rheotactic response for orientation. The fact that fish can be imprinted to synthetic substances is particularly relevant to the acquisition of knowledge by learning. It would be hard to imagine that fish had an innate ability to detect this unknown substance. If odors are environmental cues responsible for imprinting fish with information useful for subsequent homing, is it not possible to postulate many other cues such as temperature, salinity, various characteristics of currents, etc. that could also be used in a similar fashion? Whatever the cue, it must be present in the environment in a recognizable form at all times of the year or at least at the time of imprinting and a subsequent return.

Initiation of Migration

Migrations from one place to another require orientation abilities along the way and possibly a type of map in order to arrive at the final destination. For a migration to be successful it must be initiated at the appropriate time and must be terminated when the goal is reached. Thus, in addition to orientation capabilities, fish must include appropriate cues to start and stop the migration at environmentally desirable times.

There are a number of biotic as well as abiotic factors or cues which stimulate the appropriate behavior responses necessary for migration. However, these proximate stimuli are of significance only if the fish has developed a threshold physiological state. The attainment of the desired physiological or neurological states prior to initiation of migration is discussed in Chapter 5. Baker (1978) recognises two types of migrations depending on the type of initiating cues. He uses the term facultative migration to refer to a series of physiological events which lead to an initiation of migration in response to a certain value of a habitat variable or variables. Habitat variables are, for example, population density, food availability, intraspecific behavior (dominance, hierarchies, territoriality), interspecific behavior (predation, competition), that form unpredictable fluctuations in the environment. Baker describes the second type of migration as obligatory and refers to a sequence of physiological events which leads to an initiation of migration in response to an indirect variable that is predetermined during a specific stage of development. Indirect variables are such things as temperature, photoperiod, rainfall, lunar cycles, water currents, etc. that are generally predictable fluctuations in the environment. He describes other cues as migration-cost variables. These factors relate to the energy cost of migration. Such a factor may be water current, volume or depth found in a stream during a freshet that would obviously benefit downstream migrations.

Due to the tilt of the earth's axis and the ensuing seasonal fluctuations, especially in temperate zones, there exist climatic conditions that are at times favorable and at other periods unfavorable. These may provide reasons and also cues to initiation of migration. These seasonal fluctuations can thus lead to circannual cycles in fish migratory readiness and the changing environmental cues (temperature and photoperiod) may set the phase of these cycles and initiate desired behavioral migratory responses (see Meier and Fivizzani, 1980).

Temperature appears to be one of the most important proximate factors causing many different species to migrate. Temperature seems to be the major stimulus for many anadromous salmonids when they leave nursery streams (Fried *et al.*, 1978; McCleave, 1978; Solomon, 1978; Hartman *et al.*, 1982) or lakes (Hartman *et al.*, 1967) as young fry or smolts. Migration of prespawning common shiners, *Notropis cornutus* (Dodson and Young, 1977), redside shiners, *Richardsonius balteatus* (Lindsey and Northcote, 1963) and brook sticklebacks, *Culaea inconstans* (MacLean and Gee, 1971) are examples of the species diversity and stages in the life cycle that are initiated by increased temperatures as directive factors. Movement of fish in the marine environment is also correlated with changing temperatures (reviewed by Leggett, 1977). A drop in body temperature due to environmental fluctuations initiates migrations of the anchovy, *Engraulis encrasiocholus*, but curiously also stimulates the fattest fish to leave first (Tarenenko, 1966). Temperature in this case is a proximate factor that triggers those fish to migrate that have attained the desired morphological and physiological condition required for the rigors of migration.

There is relatively little evidence in support of photoperiod being an important cue in initiating migrations. For example, Atlantic salmon, *Salmo salar*, start their seaward migrations in different geographic locations at various rates when the photoperiods are quite different but at times when the water temperatures are similar (McCleave, 1978). Dodson and Young (1977) demonstrated that long photoperiods caused an increased frequency of upstream migrations of common shiners but again temperature changes appeared to be significant. What seems to be more important than the length of the photoperiod is the change from light to dark. Once an appropriate physiological or behavioral state has been acquired, many species of fish will migrate but only at certain times of the day. Many species are crepuscular in their movements and migrate at sunrise (brown trout, *Salmo trutta*; Swift, 1964), sunset (juvenile herring, *Clupea harengus*; Stickney, 1972), or both (alewife, *Alosa pseudoharengus*; Richkus, 1974). The migration of young Pacific salmon species occurs mainly at night (Hartman *et al.*, 1967, 1982; Byrne, 1968) probably to avoid predation.

There are a number of other indirect variables that have been implicated as stimuli for migration in at least a few fish species. Thus, there is evidence that fish are affected by various changes in lunar cycles. Fish might perceive such changes due to gravitational effects, visual cues or indirectly via tidal inferences. Such migratory movements are best exemplified by the California grunion, *Leuresthes tenuis* (Walker, 1949). These spawning migrations occur only at night during March to August at the time of a full or new moon when the tides reach their highest level. The eggs of the fish are buried in the sand where they stay until the next large high tide returns two weeks later. The actual component(s) of the lunar cycle that the fish use as a cue are not currently known. Such lunar or semilunar spawning migrations have been reported in other fish species as well (see Walker, 1949; Grau, 1982). The seaward movement of juvenile Pacific salmon may also be timed by lunar cycles (Mason, 1975; Grau, 1982).

There are a number of marine species which migrate up and down the intertidal zone in response to the changing of the tides (see Schwassmann, 1971). In the case of the blenny, *Blennius pholis*, these movements are in response to changes in hydrostatic pressure resulting from tidal cycles (in Meier and Fivizzani, 1980). Daily changes in the tides could be a stimulus for a type of modulated drift or selective tidal transport whereby fish hold position on the bottom or near shore during one tide and ascend in the water column to drift during the other tide. Such tide transport has been demonstrated for migration in plaice, *Pleuronectes platessa* (Greer Walker *et al.*, 1978) but there is a lack of evidence for Atlantic salmon during tracking studies as the smolts traverse estuaries on their way to the sea (Fried *et al.*, 1978; McCleave, 1978).

Stream currents are also important cues for initiating migrations. As outlined earlier, emerging sockeye salmon, rainbow trout and cutthroat trout will be either positively or negatively rheotropic depending on their genetic makeup (Kelso *et al.*, 1981; Northcote and Kelso, 1981). Seaward migrations of smolts of Atlantic salmon (Solomon, 1978), and coho salmon, *Oncorhynchus kisutch* (Hartman *et al.*, 1982), appear to be initiated by rainfall or the succeeding freshet with its associated current increase. A number of Pacific salmon species are known to exhibit a change in salinity preference at the time of migration and the osmolality of the water may thus be an initiation cue (McInerney, 1964; Hurely and Woodall, 1968). The summer migration to deeper water of ciscos, *Leucichthys artedi*, is correlated to increasing temperature but the fall return migration is correlated to declining O_2 and increasing CO_2 levels (Fry, 1937).

There is evidence that some fish species initiate facultative migrations in response to a habitat variable. In many streams coho salmon are produced in numbers greater than the stream carrying capacity. The seaward migrations are proportional to the stream density and smaller or subordinate fish appear to be displaced due to aggressive behavior of larger fish (Chapman, 1962; Mason, 1969; Hartman *et al.*, 1982). On the other hand, it is the larger sockeye salmon smolts in lakes that seem to migrate seaward first (see Hartman *et al.*, 1967). This may be due to some intraspecific behavior or may be due to a migration-cost variable since the larger fish are the only ones physically capable of making the necessary migrations.

Fish that are physiologically ready to migrate (Chapter 5) may initiate their migration in response to more than one environmental cue or may have one cue interacting with various other cues. Temperature has already been discussed as a major factor initiating different fishes' migrations but this cue has been shown to be modified by photoperiod (Byrne, 1968; Dodson and Young, 1977). Temperature affects the daily vertical migrations of herring but the response appears to be a compromise with the degree of satiation (Bitzukov, 1959). Responses of upstream migrating alewives to temperature are modified by CO_2 levels. Thus, a change in the choice of migration routes or timing could occur due to a temperature difference but this could be overbalanced by a greater change in ppm CO_2 (Collins, 1952). Migration initiating cues thus vary between fish species as well as temporally within species.

Goal Recognition

The mechanisms of orientation for a variety of fish species migrating to a specific site are well documented but the environmental cues used to recognize the goal are less well investigated. Freshwater and intertidal species appear to recognize their goal by the use of olfactory cues emanating from the home site. Studies with anosmic fish indicate that visual cues of the home site geography are also important (Gunning, 1959; Harden-Jones, 1968; Aronson, 1971; Khoo, 1974; Goff and Green, 1978). Many anadromous species seem to depend mainly on olfactory cues (see Leggett, 1977).

Orientation Mechanisms

Our knowledge of the ways and means by which fish find their way during long migrations has come largely from studies of commercially important species, especially salmonids and eels. More recently, however, mechanisms have been investigated for a wide variety of fish species encompassing many different families (see reviews by Harden-Jones, 1968; Hasler, 1971; Leggett, 1977; Able, 1980). The type of orientation behavior employed usually depends on the habitat through which the fish must travel such as stream, river, lakes, estuary, intertidal zones or open ocean. But some orientation mechanisms such as rheotropic responses are useful in many different habitats; many species of fish encounter drastically different habitats *en route* (see Chapter 2). Even in a specific environment an individual fish may have the ability to use a large repertoire of mechanisms depending on the prevailing circumstances.

Celestial Cues

Solar Orientation. There is now considerable evidence that fish can use the sun for orientation during migration (see Hasler, 1971; Able, 1980; Meier and Fivizzani, 1980). If conditions are favourable for fish to observe solar changes they may use different aspects of these changes that occur in the daily sun cycles. The fish might observe the changing angle of the sun relative to the horizontal plane, which is termed the sun's azimuth. On the other hand, the change in angle of the sun in the vertical plane, termed altitude, may give a fish useful information for orientation. Descriptions of the sun's azimuth and altitude and their changes occurring due to the time of day, season or latitude are well illustrated and described by Hasler (1971). As the sun's azimuth and altitude both change during the day, fish that use these cues for orientation must possess some time-compensating mechanism. Although the sun may be useful for orientation it does not imply that the sun is essential (Groot, 1965) and indeed fish can orient in the absence of the sun as, for example, on cloudy days (Quinn, 1980), at night (Hartman *et al*., 1967; Madison *et al*., 1972; Stasko *et al*., 1973, 1976), in covered tanks (Miles, 1968; Quinn, 1980) or after blinding (McCleave, 1967; McCleave and LaBar, 1972). Contrarily, the demonstration of some other orientation mechanism does not rule out solar orientation. Experiments to test if a fish is using a time-compensated sun-compass have been conducted to see if the fish will maintain a given compass direction at any time of the day and if this orientation could be eliminated if the sun were obscured. More definitive evidence has been obtained when the animal's internal clock is phase shifted (acclimated to a different photoperiod) and the fish displays predictable changes in orientation.

The possibility of solar orientation by fish was described initially by experiments of Hasler *et al*. (1958) whereby small floats were attached to white bass before release after displacement from their spawning grounds. On clear days the fish consistently swam towards the spawning grounds which were probably too distant for olfactory or visual recognition. The fish were reported to be disoriented on cloudy days or when they were fitted with opaque eye cups. These same workers subsequently showed that white bass, pumpkinseed sunfish, *Lepomis gibbosus*, and bluegill sunfish orientated to a compass direction when trained to an artificial sun. Since these earlier studies, a number of investigators have demonstrated sun compass orientation in a number of different fish species (Braemer, 1960; Hasler and Schwassmann, 1960; Kalmus, 1964; Schwassmann and Hasler, 1964; Groot, 1965; Schwassmann, 1967; Goodyear and Ferguson, 1969; Goodyear, 1970, 1973; Loyacano *et al*., 1977; Goodyear and Bennett, 1979). Most of the above experiments have been conducted with fish held in experiment tanks where it is difficult to assess the extent to which the fish would use the sun compass orientation if free to migrate in its natural environment. Many fish that display sun-compass orientation in the laboratory might employ other mechanisms when other appropriate cues are available in the natural habitat. In this connection, the study by Winn *et al.* (1964) is of particular interest. They studied two species of parrot fish, *Scario guacamaia* and *S. coelestinus* that migrate daily to inshore feeding areas and return each night to offshore caves. When displaced, these fish migrated in the appropriate direction from the capture site but when clouds obscured the sun, at night, or when the fish were blinded by eye caps their orientation was disturbed. The most convincing aspect of these experiments was that the fish reoriented themselves when their biological clocks were phase shifted. The degree of this reorientation was appropriate for the magnitude of the phase shift. Such experiments have been confirmed by Goodyear and Bennett (1979) using free swimming bluegill sunfish that were phase-shifted. Bluegills subjected to light regimes advanced by six hours displayed predictable changes in their preferred movement directions (15° per hour). Recently, Quinn (1980) has conducted orientation studies with lake migrating sockeye salmon fry. When tested in experimental tanks the fry orientated in compass directions appropriate to the directions required for migration during both day and night. Orientation was also maintained on cloudy days as well as when the tanks were covered with plastic. Clearly, some other mechanism besides sun-compass orientation was being employed by the fish. If the fish were exposed to a 90° counter-clockwise shift in the horizontal component of the earth's magnetic field, the mean direction of fry movement also changed by approximately 90° at night when given a view of the sky. During the day fish reoriented themselves in accordance with the altered field only if the tanks were covered or under total overcast (Quinn, 1982a). If the fish had a view of the sun but were still exposed to an altered field, their geographically appropriate compass bearing was unchanged. These experiments are interesting in that they not only demonstrate sun compass orientation but show that such an orientation is of a higher priority than another environmental cue (earth's magnetic field), at least when the fish has access to solar observations.

Sun compass orientation implies a number of capabilities for a fish. Obviously, the fish must be able to 'measure' the angle that the sun makes with either the horizontal and/or vertical plane. The fish may also possess the ability to evaluate the rate of change of such angles. If these angles (azimuth or altitude) that the sun makes as it traverses the sky are to be useful for geographical compass directions, the fish must be able to sense the time of day as well as the time of year, that is, possess a biological clock and calendar. On top of these abilities the migrating fish must have the ability to use the available information and 'calculate' the desired direction of movement. Although these abilities undoubtedly exist, and are a source of amazement to the experimenter, their nature might never be understood. There is some information available however as to the relative importance of the sun's azimuth and altitude to fish sun-compass orientation. Hasler (1971) conducted experiments with bluegill sunfish and white bass to assess the role of the sun's azimuth for orientation. Fish were trained to seek cover in a given direction and experiments were conducted at set times before and after noon. At these times the angles of deviation of the sun from its zenith were thus equal but in opposite directions; the sun's altitude was thus similar but the sun's azimuth was different. Experimental fish oriented in the appropriate direction could correctly find their goal when an artificial light source replaced the sun. These experiments demonstrate that these two species can use the sun's azimuth as a cue but at the same time do not rule out the use of the sun's altitude. Nearer the tropics the sun's azimuth changes more drastically during the day as well as with the season. Fish migrating over great distances and encountering large latitudinal changes towards the Equator may thus require further information obtainable from the sun's altitude. Experiments by Braemer and Schwassmann (1963) and Schwassmann and Hasler (1964) investigated the effect of the sun's altitude change without a change in the sun's azimuth. This was accomplished by allowing the test fish to see the sun only from a mirror reflection and as the mirror was tilted the perceived altitude would change. These experiments demonstrated that the rate of change of the sun's altitude was important in determining time compensation during sun-compass orientation in fish.

The plane of polarization of sunlight might also be useful to fish for orientation. The pioneering work of von Frisch (1949) with honeybee foraging flight behavior demonstrated that animals could sense and use polarized light for orientations. Following this original study many arthropods, one cephalopod, and certain fish, amphibians and birds have been shown to orient with polarized light (see Able, 1980). Fish might use polarized light since underwater polarization patterns are produced by primary scattering of directional light in the water. This is a function of the sun's azimuth, sun's altitude and depth of water (Waterman and Westell, 1956). Groot (1965) proposed that seaward migrating sockeye salmon smolts might orient to patterns of polarized light during the dawn and dusk peaks in migration following his studies of orientation. Subsequently, certain fish species were described that could perceive polarized light (Dill, 1971; Forward *et al.*, 1972; Davitz and McKaye, 1978). In addition, Stewart (1962) has shown that adipose eyelids (common in many pelagic fish) are strongly birefringent and could function to sense polarized light although this has been refuted by Dill (1971). Further evidence for polarotaxis in fish comes from studies with the Pacific half-beak (Hemiramphidae), *Zenarchopterus dispar*, by Waterman and Forward (1970, 1972) and Forward *et al.* (1972). They placed fish in underwater (3-5 m) aquaria with polarizing filters as covers and observed orientation as the polarization pattern was changed. Results were difficult to interpret so other experiments were conducted in shallow aquaria on land where polarotaxis was evident and appeared to dominate over sun-compass orientation.

Lunar Orientation. The night sky may also provide celestial cues to fish for determining compass direction. This author is not aware of any studies implicating stellar orientations in fish as is known for some bird species, but there is some evidence that the moon might be useful to at least one fish species. Brannon *et al.* (1981) tested orientation of newly emerged sockeye salmon in the absence of current but either in view of the night sky or under black plastic covers (Figure 3.5). The presence of the moon greatly enhanced the strength of

Figure 3.5: Compass orientation of sockeye salmon fry from Weaver Creek, British Columbia. Diagrams on the left indicate fish tested in the presence of the moon and on the right fish tested in the absence of the moon. Fry were tested at Weaver Creek and Harrison River on the night of their caputre and seven days later. Stippled areas are proportional to the numbers of fish trapped in those directions and the arrows indicate mean bearings. The presence of the moon greatly enhanced the strength of the northward orientation response.

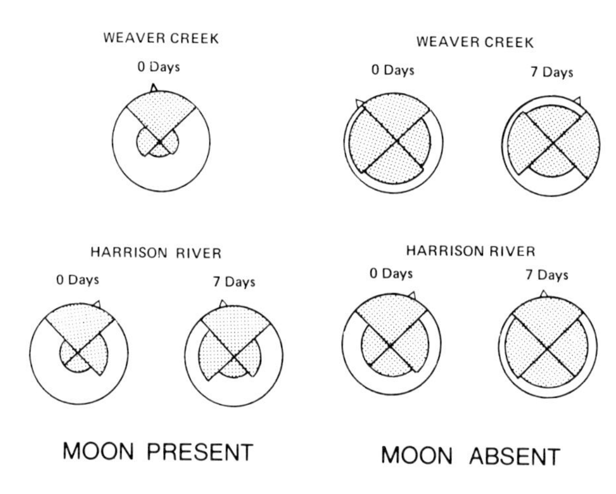

Source: redrawn from Brannon *et al.*, 1981

the orientation response. These results are puzzling since sockeye salmon at this stage in their life cycle are photonegative and migrate less during bright moonlit nights. If some fish do indeed use the moon as a cue for orientation they must possess some biological clock tuned to the different phases of the lunar cycle. The existence of a time-compensated moon compass in fish needs further investigation and may turn out to be a controversial topic as it has for the few arthropods that apparently exhibit such mechanisms (Papi and Pardi, 1963; Enright, 1972; Sotthibandhu and Baker, 1979).

Magnetic Fields

The fact that some fish can still maintain a compass direction when tested in covered experimental tanks and devoid of sensory contacts such as vision, current or olfaction suggests the existence of other possible orientation cues. The use of magnetic or electric fields for orientation has been postulated for many decades but relevant experiments have been conducted only recently (see Able, 1980). Certain fish species have the ability to produce as well as receive electrosignals (Lissmann, 1958). These fish exhibit electro-location and are so-called 'active' fish (Pals and Schoenhage, 1979). Other fish species appear to be 'passive' and can only receive such signals by electroreceptors.

Information on electro-location described for the electric fish *Gymnarchus niloticus* (Lissman, 1958) has been extended to demonstrate that some sharks can locate their prey and the electric fish, *Sternopygus* sp., can electro-orient by sensing distortions of its own electric signals (Kalmijn, 1971). Fish that can only passively sense electric fields from their natural environment might receive the appropriate cues from a variety of sources. Water currents, especially sea water, flowing through the earth's magnetic field result in an induction of an electric field that may be detected by some fish. Alternatively, the fish itself might produce an electric field within its own body as it swims through the earth's magnetic field.

Electric fields induced by water currents flowing through the earth's magnetic field are perpendicular to the direction of water flow. Naturally occurring voltage gradients have been measured in the ocean and if the fish swam parallel to the current a potential gradient perpendicular to the main axis of the fish would be generated (Rommel and McCleave, 1972). The fish might even have the ability to detect the direction of the water current flow. More recently, Pals and Schoenhage (1979) have demonstrated that there are also local electric fields caused by bottom topography and regional electric fields due to electromagnetic processes. They suggest that such patterns can provide an 'electric map' of the fish's environment. It is interesting to note that the American eel, *Anguilla rostrata*, has been demonstrated to exhibit a sensitivity to perpendicular electric fields but not to fields parallel to the fish's main body axis (McCleave *et al.*, 1971; Rommel and McCleave, 1972). However, attempts to substantiate these findings were unsuccessful (Zimmerman and McCleave, 1975). This observed sensitivity is within the range of naturally occurring gradients which also seems to be the case for some elasmobranchs (Kalmijn, 1978). European eels, *A. anguilla,* may also possess these orientation abilities (Branover *et al.*, 1971; Ovchinnikov *et al.*, 1973; Teach, 1974). Fish swimming through the earth's magnetic field could also be expected to generate fields of sufficient magnitude that could be detectable by fish species with appropriate electroreceptors (Branover *et al.*, 1971; Zimmerman and McCleave, 1975). To answer the question of whether American eels detect and use fields generated by ocean currents or by the swimming of the fish itself, McCleave and Power (1978) designed experiments using appropriate electric and magnetic field manipulations. They observed that the fish were not influenced by electric fields generated by their own swmming but orientation apparently could be influenced by electric fields generated by ocean current systems. Pacific salmon also appear to use magnetic compass orientation. Quinn (1980, 1982a) and Quinn and Brannon (1982) have conducted experiments with lake migrating sockeye salmon fry and demonstrated that a 90° counter-clockwise shift in the horizontal component of the earth's magnetic field was associated with approximately

Figure 3.6: Orientation tests of yearling sockeye salmon smolts trapped at the outlet of Babine Lake, British Columbia on their way from the lake downstream to the ocean. Each line represents the mean bearing of the smolts from one day of testing, with the length of the line directly proportional to the r statistic for that day (the radius of the circle corresponds to an r value of 0.75). The arrows on the perimeters of the circles represent mean bearings. (A) Smolts tested in a normal magnetic field and with a view of the sky. Mean bearing at 327°. (B) Smolts tested in a magnetic field rotated 90° counter-clockwise and with a view of the sky. Mean bearing of 342°. (C) Smolts tested in a normal magnetic field and under black plastic covers. Axis of bimodality 342°-162°. (D) Smolts tested in a magnetic field rotated 90° counter-clockwise and under black plastic covers. Axis of bimodality of 286°-106°. With a view of the sky, the smolts oriented towards the lake's outlet in the normal magnetic field (A) and in a field rotated 90° counter-clockwise (B). Under opaque covers, the smolts displayed a bimodal distribution. In the normal magnetic field (C), they oriented towards or away from the lake's outlet. In the altered field (D), the axis of the distribution was rotated 56° away from the axis in the normal field.

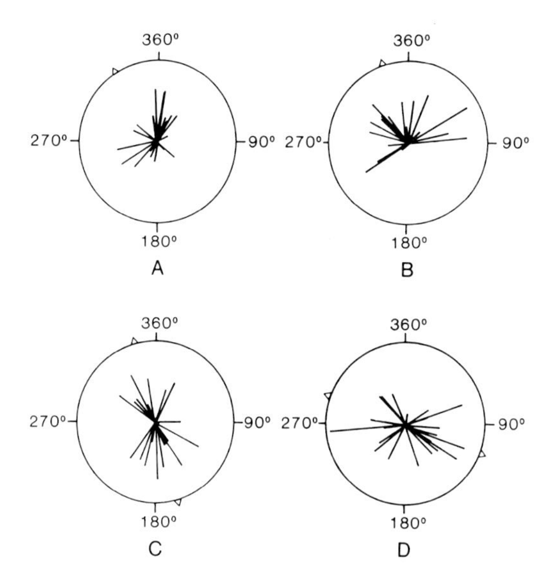

Source: redrawn from Quinn and Brannon, 1982

90° changes in the fry directional movements (Figure 3.6). This change in direction occurred both at night and during the day if the fish had covers over their tanks but not when the fish had a view of the sun. Thus, the presence of this ability to orient using a magnetic compass is only part of a repertoire of mechanisms and in this case appears to be hierarchically lower than the use of celestial cues. Brannon *et al.* (1981) have also tested sockeye salmon fry for orientation to a directionally altered magnetic field but found little or no response. However, as these authors point out, the migratory situations for the fish they tested were quite different and the use of a lake-appropriate magnetic compass orientation would tend to retard movement downstream and even concentrate fry in unsuitable feeding areas. Open ocean migration of Pacific salmon would be likely to benefit more from magnetic compass orientation and tests of fish from this environment would be most interesting.

The use of a magnetic compass orientation mechanism implies that the fish must have the appropriate sensory apparatus. Able (1980) has outlined some general models of magnetic detection which involve: (1) induction of a measurable current when an organism moves through the earth's magnetic field; (2) some sort of paramagnetic material which responds in a magnetic field, or (3) deposits of permanently magnetic material in the animal that detect the external field in some manner. The honeybee has been shown to possess magnetic material in its abdomen that could serve as a transducer (Gould *et al.*, 1978). Whether fish possess such materials is not yet known but some species do seem to be able to detect induced currents that may arise from movement through the earth's magnetic field. The elasmobranch ampullae of Lorenzini has sufficient electrosensitivity to detect oceanic electric fields (Dijkgraaf and Kalmijn, 1962; Kalmijn, 1966). Pit organs or ampullary organs in catfish, Gymnotids and Mormyrids may also be sites that are useful to these fish for electro-location (Bennett, 1971). The fact that some elasmobranchs have electroreceptors that could possibly function for compasss orientation does not mean that they use them for this purpose.

Walker *et al.* (1982) have recently been able to locate magnetite in the ephmoid bone of the yellowfin tuna, *Thunnus albacares*. These deposits are centrally located in the fish and may prove to be the site of a goemagnetic field sensory apparatus as these fish could be conditioned to discriminate between different Earth-strength magnetic fields.

Currents

It is not difficult to imagine that fish may take advantage of currents as cues for orientation since they usually have visual, tactile or inertial stimuli from their environment as well. Once a current has been detected, a fish may orient to the current and actively swim with or against it or be passively carried along. The use of currents in the open ocean is probably more difficult to appreciate. Even in this latter environment, fish may be able to perceive currents by their lateral line senses or other organs that, in conjunction with other means, could give the animal useful information for orientation. Currents in large water masses have characteristic features, especially at interfacial or boundary zones. Such characteristics could be irregular turbulence, temperature differences, salinity gradients, turbidity changes, etc. that could give the fish visual or tactile cues as in rivers. One might expect fish to gather information from currents that exist in the stream and ocean environments but useful information could also be obtained from currents found in tidal zones or displacements of water found to occur due to wave action. Diadromous species of fish that must cross estuarine and intertidal areas might well benefit by appropriate rheotropic responses to tidal currents. Many areas of continental shelves receive consistent and regular patterns of wave action that in turn set up disturbances or wave surges on the bottom. This action may also be useful to some fish for orientation. Currents are usually thought of as being capable of transporting adult fish or at least providing cues to which adult fish may respond for desirable rheotropism. Nevertheless, such currents may also be very useful to movements of fish eggs or young larvae. Such movements of larvae were thought to be passive but there is evidence that some fish larvae may actually regulate the rate and direction of their migrations by rheotropic responses (see Leggett, 1977).

Experiments with rainbow trout from above-waterfall stocks (Northcote and Kelso, 1981) or from inlet and outlet stream stocks (Kelso *et al.*, 1981) have clearly demonstrated an innate rheotropic response by young fry. Such responses to current have been observed for many other freshwater species as well (Rayleigh, 1967, 1971; Rayleigh and Chapman, 1971; Brannon, 1967, 1972; Arnold, 1974). The response of fish to currents in a riverine environment depends not only on innate rheotropic responses but also on modifications by various other environmental parameters. Rheotropism causes fish to orient to a current as well as to adjust their swimming speeds in response to the rate of flow of the current. Dodson and Young (1977) have demonstrated that rheotropic behavior of migratory common shiners could be modified by temperature and photoperiod. They suggest that such regulation of rheotropism would be important for such an upstream migrant to ensure arrival at the spawning grounds during optimal environmental conditions. Rheotropic responses of rainbow trout also appear to be affected by temperature and illumination (Kelso *et al.*, 1981). Other environmental factors have also been implicated in modifications of fish rheotropic responses. An enhanced rheotaxis has been observed in rainbow trout in response to decreased prey abundance both in field (Slaney, 1972) as well as laboratory experiments (Slaney and Northcote, 1974). Emanuel and Dodson (1979) have shown that currents may provide directional cues for migrating adult male rainbow trout and that olfactory stimuli from females' ovarian fluid regulate the rheotropic response.

The response of fish to currents in the open ocean is less well understood due to the obvious difficulties of access and experimentation. The reviews by Royce *et al.* (1968) and Banks (1969) suggest that migrations of salmon in the northern Pacific Ocean are active, directional and follow patterns of current flow. Leggett (1977) reviews historically documented changes in movements of herring and Greenland cod associated with known oceanic current fluctuations. Barber (1979) speculates that Pacific salmon cue to the consistent patterns of currents in the upper layer of the ocean, the so-called Langmuir circulations, for continuous guidance.

Intertidal species of fish and especially migrating diadromous species that traverse estuarine environments twice in their life cycle, appear to use tidal fluctuations as directional cues. Ultrasonic tracking studies of mature Pacific salmon indicate that individual fish move in unpredictable patterns which are extremely variable (Stasko *et al.*, 1973, 1976; Groot *et al.*, 1975). Tidal currents did relate, however, to the fish's movements in that salmon would often drift passively with the tidal currents and at other times would actively follow axes of tidal currents. The fish nevertheless appeared to sense and react to the presence of tidal ebbing and flooding. Seaward migrating fish could effect a selective tidal transport if they could drift with an ebb tide and actively stem a flood tide. Ultrasonic tracking studies of Atlantic salmon smolts however showed no such indications and the fish appeared to reach ocean waters mainly by passive drift (Fried *et al.*, 1978; McCleave, 1978). Contrary to the results for migrating Atlantic salmon, tracking studies of European eel elvers (Creutzberg, 1961) and adult American shad, *Alosa sapidissima* (Dodson and Leggett, 1973) exhibited selective tidal transport when they entered fresh water. When the tide was ebbing the fish swam actively against the current or when flooding they rose from the bottom and either swam with or were carried passively by the current. Such orientation to tidal currents has also been demonstrated by tracking studies of fish migrating in the open ocean. Plaice (Greer Walker *et al.*, 1978) and Atlantic cod, *Gadus morhua* (Arnold *et al.*, personal communication cited in Fried *et al.*, 1978) both change their depth in the water in accordance to the tidal cycle.

Olfaction

The use of olfactory cues was probably the first proposed mechanism for fish migrating to their home stream (Buckland, 1880). Preliminary supportive experimental evidence was only gathered, however, when Craigie (1926) conducted studies on homing adult sockeye salmon. Again a hiatus occurred until Hasler and his co-workers revived and conducted appropriate experiments on the so-called olfactory hypothesis. This stated that fish learn distinctive odors of the home stream during some early period of residence and the odors are later used as a cue for the adult spawning migration. These studies originated by demonstrating that the bluntnose minnow, *Hyborhynchus notatus*, could discriminate between water rinses from various aquatic plants (Walker and Hasler, 1949). Subsequent experiments indicated that this fish could detect differences in water from two different stream waters (Hasler and Wisby, 1951). Later work by Hasler and others accumulated data to support the olfactory hypothesis (Hasler, 1954, 1956, 1960, 1966, 1967, 1983; Scholz *et al.*, 1972, 1975, 1976; Johnsen and Hasler, 1980). More recently, there has been a large amount of information made available which also adds credence to this hypothesis (see reviews Harden Jones, 1968; Hasler, 1971, 1983; Woodhead, 1975; Leggett, 1977; Baker, 1978; Meier and Fivizzani, 1980). Olfactory cues have thus been shown to be important to upstream migrants but the extent to which these cues are useful for fish in the marine environment is more speculative. The distance into the ocean that these freshwater-borne odors reach, however, may be of great significance. There is some evidence that cues from natal streams do reach fish in the ocean and in some instances possibly travel many tens of kilometers away from the river mouth (see Hasler, 1971). Short distance migration or homing behavior by oceanodromous species has received some attention recently

Figure 3.7: Orientation to a home site of normal, blind, anosmic, and unilaterally blind and anosmic (unlabelled) radiated shannies. The small arrows indicate the mean direction chosen by each group. The data indicate that olfactory contact with the home site is involved in the steering mechanism. The home-site fidelity of anosmic fish, and the fact that some anosmic fish homed, indicated that vision may also be involved in the recognition of a restricted area around the home site.

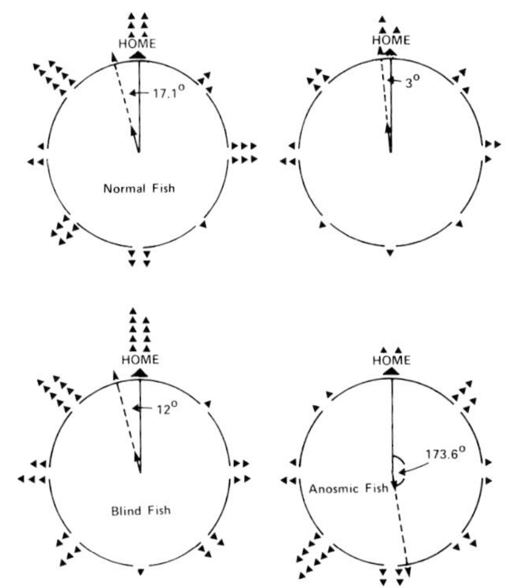

Source: redrawn from Goff and Green, 1978

and there is evidence that olfactory cues are important for these species as well (Khoo, 1974; Goff and Green, 1978) (Figure 3.7).

Barber (1979) has made a rather interesting hypothesis regarding olfactory cues in the open ocean that may assist migrating young salmon. He speculates that since wind and wave patterns over the North Pacific Ocean are relatively persistent, the resulting Langmuir circulations also give reliable current cues that indeed may contain useful odorous compounds left there by preceding adult fish populations.

If the fish use olfactory cues to recognize their natal stream and especially if this is possible at some distance from the river mouth, they must possess an acute sense of smell. Fish are known to respond to olfactory cues originating from their home stream and to test the extent to which Atlantic salmon could detect such odors, Sutterline and Gray (1973) found that these fish could still discriminate water that had a content of only 0.1 per cent home stream water added. Some fish react to water rinses from mammalian (human, bear or sea lion) skin (Alderdice *et al.*, 1954; Brett and MacKinnon, 1954). The active substance in this case was found to be L-serine which can elicit reactions at dilutions up to 1.3×10^{-10} (Idler *et al.*, 1956, 1961). Hara (1970) has reviewed investigations carried out on a wide variety of fish species and covering many different organic compounds to ascertain the sensitivity of fish olfactory capabilities. One notable study was conducted by Teichmann (1962) employing trained European eels. He tested a number of different substances and one (β-phenylethyl alcohol) had a threshold of detection at a dilution of 3.5×10^{-19}.

He made estimates that only a few molecules were required in the olfactory sac to elicit an olfactory response electrophysiologically. Such studies certainly add credence to the sensitivity that fish would require to orient according to the olfactory hypothesis.

A number of researchers have attempted to ascertain the nature of the odorous compound(s) present in the homestream environment that homing fish use as a cue. Hasler (1971) has outlined a set of criteria that such a compound must exhibit for use as a stimulus. It must remain constant over many years as fish may be absent from the home stream for this length of time. Changes in the factor(s) must not occur faster than the rate of evolution of the particular species. The substance(s) should not be cyclical since fish may return during different times of any one year. The compound(s) must be specific to only the salmon that home to a particular area and not generally induce migration for the whole species. Finally, the factor(s) must remain detectable even in the event of drastic physical, chemical or biological perturbations while the fish is away at sea. Employing various separation techniques, experimenters have described the factor as being organic, inorganic, volatile and non-volatile, heat-labile, water-soluble, acidic, basic, and dialyzable (see Leggett, 1977). These differences might not be surprising since different streams are likely to have variations in chemicals that make them unique and recognizable to the fish populations that home there. There have been suggestions that the factor(s) may emanate from home stream aquatic organisms and soil constituents. One possible source of odors that has received attention recently is from conspecifics, that is, pheromones. Char, *Salvelinus alpinus*, chinook and coho salmon can detect water in which they themselves, conspecifics from other populations or other species have been held (Nordeng, 1971; Dijon *et al*., 1973b; Døving *et al*., 1974). Some streams, previously devoid of upstream adult migrating Atlantic salmon, had fish enter only after previous stocking with young fish (White, 1934; Soloman, 1973). The source of such pheromones has been described as coming from skin mucous (Nordeng, 1971; Døving *et al*., 1973), intestinal contents (possibly bile) (Selset and Døving, 1980) and ovarian fluid (Emanuel and Dodson, 1979). To be useful for homing, the pheromones must be population-specific and there is little or no evidence that this is the case for salmonids. Knowledge of the nature of these compounds will be most useful in understanding how olfactory imprinting might occur and would be beneficial for managing new as well as established fish spawning areas.

Artificial imprinting by the use of decoy odors has given substantive support to the olfactory hypothesis. It has been possible to imprint coho salmon to a synthetic compound (morpholine) and to demonstrate that the fish can retain this information and use it to find the home stream (Cooper and Hasler, 1976; Cooper *et al.,* 1976, Johnsen and Hasler, 1980; Hasler, 1983). These studies show that the imprinting compound need not necessarily be a pheromone or for that matter any other naturally occurring compound. These experiments have employed direct observations or ultrasonic tracking techniques to show further that the salmon exhibit appropriate behavioral responses when they encounter the decoy odor. Johnsen and Hasler (1980) demonstrated further that fish imprinted to morpholine exhibited a positive rheotaxis whereas fish not imprinted to morpholine showed a negative rheotaxis. It appears that imprinting odors are thus a stimulus for fish to proceed in their migration by using an appropriate orientation and behavior. Differential responses of fish to currents occur when the fish are imprinted to different olfactory compounds that may be the basis for segregation of fish to various homestreams. If specific compounds can cause homestream imprinting in fish, serious consideration should be directed towards substances found in industrial effluents that may thus disrupt spawning migrations. If fish do imprint to such compounds it is not difficult to envisage such substances changing in effluents, disappearing or being displaced to other sites, thus causing inappropriate responses by migrating fish.

Electrophysiological (electroencephalogram, EEG) studies have shown that fish can detect odors (Hara *et al.*, 1965; Ueda *et al.*, 1967). The use of an EEG response has been used to test whether fish, physiologically at least, can detect water from their natal stream as compared to other waters. Attempts have been made to relate the amplitude of the EEG response to an indication that the fish is detecting homestream water (Cooper and Hasler 1973, 1974; Dizon *et al.*, 1973a). However, some studies have indicated that water from non-homestream sources can elicit responses as large as or larger than natal water (Oshima *et al.*, 1969; Ueda *et al.*, 1971; Dizon *et al.*, 1973b). It appears that the EEG can only show that the fish is sensing some odorous substances. Bodznick (1975) could not find any relationship between the relative EEG responses of sockeye salmon to water and the actual behavioral responses of the fish in these same waters during migration. Bodznick, like other investigators, also found that the relative EEG varied with the same stimulant and may reflect the physiological state of the fish such as sexual maturation (Suterlin and Suterlin, 1971; Cooper and Hasler, 1973; Dizon *et al.*, 1973b). The EEG response may thus be a useful technique to investigate when or the threshold at which, a fish senses natal stream water. Its specificity and usefulness for predicting subsequent behavior is, however, questionable.

Visual Cues

Visual input has already been discussed with respect to the use of polarized light and celestial cue perception for orientation but pattern and form vision may also be useful for direct detection of underwater landmarks. There are few investigations that have studied vision during an actual migration. However, once a migration is complete the home site is often chosen or recognized by visual cues. Some marine radiated shanny display a home-site fidelity even if their olfactory rosettes are cauterized, thus indicating a reliance on visual cues (Goff and Green, 1978) (Figure 3.7). Likewise, a reduction in home-site fidelity was observed in blinded or anosmic tide-pool sculpins but there was no difference between these two groups of fish which would indicate that vision is equally important to olfaction (Khoo, 1974). Aronson (1951, 1971) suggested that the basis of orientation in the Gobiid fish, *Bathygobius soporator*, in jumping from one pool to another at low tide, is derived from a prior knowledge of the topography which was gained by visual stimulation while swimming over the pools at high tide. Freshwater species such as the green sunfish, *Lepomis cyanellus,* and the European minnow also appear to recognize their home range using visual cues (Hasler, 1966).

There is some evidence that fish might use visual cues to locate their homestream. Groves *et al.* (1968) displaced adult Chinook salmon, *Oncorhynchus tshawytscha*, 19 km upstream and downstream from their spawning site on the Columbia River. Approximately half as many blinded fish returned as intact controls. These blinded fish were in poorer condition and their returns were delayed. Very few olfactory occluded fish returned, indicating that olfaction was more important than vision for returning adults to locate their specific spawning site. Dodson and Leggett (1974) investigated the roles of olfaction and vision for adult American shad returning to the Connecticut River. Fish were sensory impaired, released adjacent to or 10 km away from the river-mouth and followed by ultrasonic tracking. Anosmic fish homed less successfully than intact shad. Nevertheless, four of six anosmic fish still located the river, indicating that visual clues may be important in maintaining a directional bias. Blind shad could locate the river when placed 10 km away but could not do so when released adjacent to the river. Visual clues for these fish could thus be important at different phases of the migration.

Open ocean migration may also require visual contact with the environment for successful migrations. Landmarks would of course be of no value to such fish but areas of ocean fronts may provide an appropriate visual stimulus for orientation. Ocean fronts are known to be areas where temperature, salinity and current differences occur. If the fronts have these differences as well as turbidity or organic material variations their visual detection could be possible.

Salinity

Diadromous fish species pass between fresh and sea water twice in their life cycle. If such fish could detect differences in salinity a useful cue might be available for an orientation mechanism. McInerney (1964) has shown that some Pacific salmon can detect differences in NaCl concentrations as low as 0.0005 M. Such abilities can be envisaged as being useful in traversing estuarine environments or possibly following distinctive ocean fronts. Fujii (1975) has correlated movements of adult sockeye salmon with known changing oceanographic characteristics. These salmon congregate just south of the Aleutians in May and seem to be restricted from entering the Bering Sea due to a low temperature/high salinity barrier in the Aleutian passes. In June this barrier is reduced by the Alaskan stream and the salmon enter the Bering Sea at this time. They then follow lower salinity/higher temperature water near the halocline which leads them into Bristol Bay where their spawning streams enter the sea.

Five of the species of Pacific salmon are known to possess different salinity preferences in accordance with the time of seaward migrations (McInerney, 1964; Hurley and Woodall, 1968). The development of the appropriate physiological state to desire a higher salinity coupled with the ability to detect slight differences in salinity could easily lead to migration and orientation towards the sea. Salinity preference has also been shown to change in the gulf killifish, *Fundulus grandis* (Spieler *et al.,* 1976b) and the three-spined stickleback, *Gasterosteus aculeatus* (Baggerman, 1959) and at an appropriate time of the year.

Temperature

Temperature has already been discussed as one of the major environmental factors responsible for initiating various fish migrations. Temperature probably affects fish during migration as well. Large bodies of water have seasonally changing thermoclines that may directly or indirectly influence fish orientation mechanisms. There is now a large body of information available that correlates fish movements with known water temperature patterns (see Leggett, 1977). The movements of cod and capelin, *Mallotus villosus* (Kondo *et al*., 1963) as well as American shad may be regulated by seasonal movements of specific water regimes. Although this phenomenon may be construed as a timing mechanism for optimum survival of adults and/or young, other species such as the albacore tuna, *Thunnus genno,* follow specific isotherms in their migrations (Alverson, 1961; Clemens, 1963; Laevastu and Rosa, 1963). In the latter case the tuna are presumably cueing to temperature regimes for orientation.

Although temperature may be a direct orientation cue for migration it may also affect fish indirectly by providing enhanced food sources or by increasing swim speeds. The vertical diel migrations of herring in the Baltic Sea seem to be a balance between temperature and food availability leading to satiation (Bitzukov, 1959). Seasonal migrations of herring are also affected by increased phytoplankton and zooplankton blooms in the Norwegian and Greenland Seas which occur when there is an inflow of warm Atlantic water (Pavshtiks, 1959). Patterns of migration are also correlated with temperature/food availability relationships in other marine teleosts (Laevastu and Hela, 1970; Leggett and Whitney, 1972; Berenbeim *et al.,* 1973). Metabolic scope for activity and prolonged swimming speeds have both been shown to increase for a variety of fish species in relation to elevated environmental temperatures up to a maximum after which there is a decline at the upper thermal ranges of tolerance (see Beamish, 1978). In this respect temperature can be viewed as a cue regulating internal processes that indirectly affect migration orientation.

Inertial Guidance

It has been postulated that animals might find their way by inertial navigation which implies that they can detect as well as integrate all the accelerations experienced during an outward-bound journey (Barlow, 1964). This type of return migration was postulated not only as a reversal of each turn experienced during the displacement but a computation to vector the mean direction of displacement and thus calculate the most direct route home. This hypothesis has been applied to fish since the goldfish, *Carassius auratus*, was shown to possibly exhibit a cumulative zero difference in degrees after making numerous left and right turns (Kleerekoper *et al.*, 1969). Pacific salmon (Hoar, 1956) as well as goldfish (Harden-Jones, 1957) can maintain such compensatory angular displacements for considerable periods of time. Harden-Jones (1968) has suggested that the labyrinth of the inner ear might be involved in maintaining this behavior. This hypothesis has been tested in birds; there is a decisive lack of support that this type of orientation exists, at least in different avian species (see Keeton, 1974).

Random Search

Studies of fish migrations, from the point of initiation to the final destination site, have led to the compilation of various species-specific characteristics. For some fish populations such parameters as average swimming speed, percentage of individuals successfully reaching the final destination, time required to traverse the migratory route, degree of wandering, precision of direction, etc. are well described. The availability of such parameters has been a stimulus to construct mathematical models to predict the nature of fish migrations. A number of such models have suggested that fish migrate to their home site largely by random search (Saila, 1961; Saila and Shappy, 1963; Patten, 1964; Harden-Jones, 1968).

Ultrasonic tracking studies conducted for a large variety of migrating fish species have been interpreted as lending support to the random search hypothesis. Many of these studies have indicated that movements of certain individual fish are not goal-oriented or are non-linear and thus appear to be random in nature. However, there are two major difficulties with this interpretation. First, out of necessity, these studies observe only the particular fish for a limited time span and thus represent only a portion of the entire migration. There may be times during a migration when movements are indeed random. These might be considered as 'side trips' from the main migration and be examples of different behavioral patterns brought about by such things as feeding activity, predator avoidance, schooling activity, etc. Secondly, the movement patterns may only appear random due to the lack of awareness of the researcher of the cues actually being used by the fish. Tracking studies show only the location of the fish in time but generally do not provide information concerning the physical, chemical and biological environment that the fish is in. The fish in fact may be cueing to some environmental parameter that, on the surface, is not readily obvious to the experimenter.

There are also problems with other arguments that are put forward in support of fish migrations resulting purely from random search behavior. Proponents of the hypothesis suggest that no convincing evidence exists that fish in nature display navigational or even orientation mechanisms. It is true that most orientation experiments have been conducted under controlled experimental situations or that evidence in support of orientation is only circumstantial as correlations of fish movement with known environmental parameters at the time. This however, does not seem to be an argument in support of random search behavior, but rather a description of the stage our understanding has reached with respect to fish orientation mechanisms. Some migratory fish wander and arrive at places other than their considered home site. There are data in support of this statement but the percentage of the total population that wander is far less than one would predict if the fish were migrating due to purely random search (see Leggett, 1977). Also, patterns of spatial scattering of fish during migration do not fit models of fish movements based solely on random search.

The model proposed by Saila and Shappy (1963) to explain the homing ability of Pacific salmon from the sea by purely random searching would predict zero returns. According to these authors only a very slight homeward bias apparently would be enough to allow the salmon to return in the desirable time and in numbers sufficient to represent actual observed returns. This slight homeward bias to the model lends support to the requirement for orientational abilities for migrating salmon. In this case the directional finding ability becomes a matter of degree. Random-search models would have to account for the fate of individuals that failed to reach the home site. As already mentioned, wandering individuals do not account for these differences. In the case of anadromous salmon, according to such models a large percentage would be lost at sea. However, these fish do not show up in commercial fisheries when others arrive at the home site. The possibility exists that those left in the sea could die as do spawning adults or that they could be lost to predators. No such evidence exists, however, for these possibilities. The model proposed by Patten (1964) has less constraints than others and allows for more interactions of the fish with its environment. During particular phases of a migration fish can use appropriate cues for orientation and then might shift to a random search for the next series of cues to continue with the next portion of the homeward journey.

Maximization of Comfort

It would be superfluous to go into great detail in this section since the subject has been excellently reviewed by Leggett (1977). The model proposed by Patten (1964) has been expanded to provide a generalized model for fish migration (Balchen, 1976a, 1976b). Balchen postulates that fish move from place to place to optimize their 'state of comfort'. Maximization of comfort is thus an unconscious search for an environment that at the time best suits the physiological state of the fish. As the biochemical or physiological state of the fish changes during its life cycle, the search for an appropriate or complementary environment will also change. Although fish can regulate their physiology to many changing aspects of their environment, there are many examples whereby a fish changes its physiology first and then enters a new environment – a type of pre-adaption (see Chapter 5). Seasonal, annual or once in a lifetime changes occur in fish that are controlled by changing external and internal environmental parameters. Changes in sensory perceptibilities and thus orientational capabilities are no exception (see Woodhead, 1975). Thus, depending on the physiological state of the fish, the initiation of migration, routes followed, cues perceived and used will be determined.

As with any other mechanisms, there are a number of factors that impinge on orientation at any one time during migration. For this mechanism to operate, the fish must therefore choose the appropriate environment and thereby compromise between relatively important variables. The fish could then perhaps make unconscious decisions at appropriate times in its life cycle as to desirable environments as assessed for temperature, salinity, food availability, dissolved oxygen concentrations, specific orientation cues, etc. This type of optimization behavior might explain periods of directed orientation movements as well as other times when fish seem to be moving in a random fashion. However, how could fish select the best conditions? At any given time they are in water with a particular temperature, dissolved oxygen, salinity, turbidity, etc. Do they swim faster or slower as conditions get better or worse? Do they turn more or less often? Do they remember recently experienced conditions and somehow compare them to their present environment? These questions need to be clarified if comfort maximization is to be a viable hypothesis explaining certain fish movement patterns.

Integration of Orientation Cues

If a migrating fish can orient without a particular cue being available it does not mean that the fish cannot or would not use that cue if it were available. Likewise, if that particular cue is used it does not mean another cue cannot be used if the first were eliminated. Such situations make experimentation on fish orientation difficult. Whether the fish is attempting to maximize comfort, which cues are important, at what time, and how are they related to each other are a number of questions for which experimental design becomes increasingly onerous.

As fish migrate from one area to another along their migration route they encounter new environmental parameters useful as cues for orientation mechanisms. Thus, depending on the availability of such cues or the physiology of the fish, certain cues will be used at different times during a migration. Returning adult anadromous species may thus use celestial cues while orienting in the open ocean but change to olfactory signs when approaching their natal stream. As with other vertebrates, fish seem to have a hierarchy of orientation mechanisms. For example, certain species of migratory fish appear to use celestial cues for orientation but when such stimuli are not pesent such as at night or during cloudy days they can switch to other mechanisms such as electromagnetic orientation. There also appears to be a sequence of cues required by certain fish species. One environmental parameter may be a cue itself but may also stimulate a response to another different parameter. Olfactory cues are thus known to stimulate appropriate rheotactic responses in migrating young Pacific salmon. In spite of such difficulties of varied and changing mechanisms of orientation, even within given individuals, recent research is answering many of the previous mysteries of animal orientation. The future is aptly described by Able (1980) who states that 'it is the very difficulty of prying loose the answers to questions about orientation and navigation that makes the research so challenging, at times frustrating, and always exciting'.

4 BIOENERGETICS

Introduction

The movement of fish from one place to another during migration requires a specific amount of organic fuel for the many metabolic requirements. Some fish provide themselves with adequate bioenergetic requirements before they initiate their migration. These fish can acquire food in sufficient quantities so that they have an amount in excess of substances required for general body maintenance. This acquisition of food is a result of appropriate behavioral patterns which may be brought about by physiological alterations. These fish, in addition, possess anatomical and physiological adaptations to store the ingested food so that it can be utilized at a later time when energy is required for migration.

Some other fish species solve the problem of providing energy-rich compounds for migration by feeding *en route*. This method too requires the necessary behavioral and physiological adaptations to provide the bioenergetic requirements for migration. Many of the fish species in this latter type of migrant require a number of different foraging strategies since they often enter vastly different environments that provide different prey items. These fish must therefore be capable of changing their food gathering techniques as well as possibly altering their metabolic processes due to a change in diet.

High energy-containing compounds required for migration are gathered and stored prior to migration by some fish species, whereas others have evolved species characteristics which allow them to meet bioenergetic requirements directly while feeding *en route*. Such characteristics to a large extent determine which strategy a fish will be capable of employing to provide the fuel for migration. However, environmental factors and the type of migration will also impinge on the way in which bioenergetic demands are met. Thus, if a migrating fish encounters waters differing in such factors as temperature, salinity or oxygen concentrations, the amount of, and rate at which energy-rich compounds must be supplied will differ. These variations may be factors determining whether or not a fish must store these compounds ahead of time or can obtain them during the migration itself. As an example it is not difficult to imagine that a diadromous fish species requires an additional energy output when it encounters water of a vastly different salinity. It has been estimated that up to 29 per cent of the total oxygen consumption is required for osmoregulation in *Tilapia nilotica* (Farmer and Beamish, 1969). The development of anatomical and/or physiological controls as well as mechanisms for carrying out osmoregulatory functions need to be considered in the light of their bioenergetic demands. Some fish species migrate within waters of relatively constant characteristics. However, these fish also may require vastly different bioenergetic adaptations due to the length of the migration. When the distance covered by a migrant increases, then the energy cost by necessity also increases. The time required to traverse this greater distance will of course also increase if the fish does not change its rate of travel. The increased energy cost of a longer migration will thus determine the amount and manner in which fuel is gathered and/or stored. The element of time may also affect the fish bioenergetically since temperate zone migratory fish species are often confined temporarily due to seasonal variations in the environment.

The energy requirements of fish for migratory activity can be met from a variety of sources. High energy organic compounds are obtainable directly as food from the environment or indirectly from storage sites within the fish's own body. There are, however, differences between and within species with regard to acquisition and/or utilization of lipids, proteins and carbohydrates, thus leading to a differential availability of an energy supply. The migrating fish may therefore be provided with an increased energy source not only by increasing food intake but also by liberating more energy from this food by possessing attributes which increase food conversion efficiency, intestinal absorption or storage facilitation. Energy may also be saved prior to, or during, particular phases of a migration by changing other mechanisms whereby such things as basic metabolic rates or activity patterns might be lowered.

The acquisition of an energy source for migration is required for direct as well as indirect energy costs. The major expenditure of energy for fish migrations is for locomotory activity. A large proportion of the acquired energy-yielding substrates are used in muscle contraction for locomotion (Brett and Groves, 1979). When muscle tissue increases its output for physical activity there is a requirement for more oxygen to be supplied for oxidative phosphorylation and an increased supply of metabolites required to produce the compounds leading up to the electron transport system. Thus, in addition to the increased energy utilization by muscle tissue itself there is a concomitant rise in energy demands by the respiratory and circulatory systems in order to acquire and deliver the needed increased demands for oxygen and high energy organic compounds (Jones and Randall, 1978).

Many fish migrations are directed towards the appropriate spawning grounds. Often these fish undergo gonadal recrudescence while undertaking their spawning migration. All fish require an added energy input above that necessary for general body maintenance in order to achieve this gonadal development. However, when this sexual development coincides with the time of a spawning migration with its energy demands for locomotion, an additional energetic demand is placed upon the fish. The additional energy requirements are not only for morphological and physiological changes in the gonads themselves but also for the development of structural secondary sex characteristics and sexual behavior manifestations. These increased demands can often be considerable. When they occur in migratory species such as Pacific salmon, that already require tremendous reserves for swimming long distances through different salinity environments and past difficult upstream obstacles, bioenergetic demands must be maximal. If such fish do not feed along the migratory route, premigratory food acquisition and storage in energy deposits of the body of the fish become of prime importance. In addition to the energetic demands for locomotory activity and gonadal recrudescence, fish require other specialized physiological mechanisms to complete a migration successfully. These mechanisms include osmo(iono)regulation, sensory perception during orientation and specialized behavioral alterations, each of which requires additional expenditures of energy. The amount of energy required for some of these processes may seem to be almost inconsequential, but when added to the total energy requirements for migration they may become extremely important and in fact may cause the fish to be pushed past its tolerable limits of endurance. On the other hand, some of these additional physiological demands in themselves may require significant energetic demands. For example, the energetic cost of osmoregulation for actively swimming rainbow trout, *Salmo gairdneri*, in fresh water may be as high as 30 per cent of the total metabolism of the fish (Rao, 1968). The problems encountered by fish to supply the energy needed for migration involve not only specialised adaptation to acquire and store energy-rich compounds but also include specialized adaptations of biochemical pathways to meet bioenergetic demands (see Driedzic and Hochachka, 1978; Mommsen *et al*., 1980).

During migration the energy demands on fish may be so substantial that the fish cannot recover from such drastic reductions in body reserves. Such post-spawning deaths occur following migrations of Pacific salmon. Although the sequence of events and final cause of death in such salmon are not well understood, the expenditure of energy and subsequent depletion of body reserves probably contributes to it. However, some races of Pacific salmon migrate only relatively short distances and after spawning appear not to be energetically depleted yet still die. Clearly, in these fish other factors besides bioenergetic exhaustion lead to their death. These other factors and their genetic compounds are unknown at this time. Even in these extreme cases of post-spawning death, the next generation still has the advantage of embryological development and early life histories in a suitable environment. Other migratory species, or even the downstream migrations of Pacific salmon, often expend considerable amounts of energy, but in the final analysis gain the advantage of a new environment. The depletion of body reserves during migration can be overcome by some fish and in addition, the increased availability of food can surpass requirements for general body maintenance and be used for body growth, gonadal maturation or be stored as reserves for the next migration.

Energy Sources

The energy source for the general activity of fish as well as for specialized functions such as migration comes from the ingested food and thus the mechanisms for assimilation, storage, mobilization and catabolism depend on the food acquisition characteristics of each individual species. The majority of fish species are carnivorous and consequently depend on, and are physiologically adapted to, an intake of large proportions of protein. There are also varying proportions of lipids and carbohydrates as well, depending on the prey species eaten.

Once ingested, the proportions of the different organic compounds may change since catabolic and storage processes vary between species with respect to the various food constituents. For instance, fish can readily mobilize proteins as a source of energy and thus they are a preferential fuel for general maintenance as well as for periods of high activity (see Moon, 1983). On the other hand carbohydrates are not stored in large quantities by fish as energy-rich compounds unlike mammals. Thus, ingestion of relatively high concentrations of carbohydrates will often lead to their catabolism shortly after absorption or conversion to other organic compounds for storage. Depending on use or interconversions, fish thus use lipids and proteins as storage products in different ratios to those originally ingested.

When fish require fuel from stored reserves there is also differential use of organic compounds. This is also affected by other factors such as concurrent feeding, type of activity or type of tissue under consideration. Muscle tissue requires glycogen as an energy source and if these stores are depleted then lipids and proteins will be mobilized and catabolized (see Driedzic and Hochachka, 1978). During sustained muscle activity lipids are known to be mobilized but under conditions of food deprivation proteins appear to be preferentially mobilized (see Cho *et al.*, 1982). If fish ingest energy-rich compounds in quantities in excess of demands needed for general body maintenance, material is available for growth of various tissues. These tissues may increase in size for general body growth such as muscle and bone tissue and may also include tissues that develop for specific functions such as gonadal structures or for storage and future utilization such as depositions in storage sites. The latter often require conversion to compounds appropriate for storage and subsequent alteration so that they can be utilized for future processes. Thus, preferentially stored compounds such as lipids and proteins must be altered in order to enter the glycolytic pathway and/or Kreb's cycle.

Since carbohydrates are the primary fuel for activity patterns such as migration, mechanisms are required by fish for lipolysis and gluconeogenesis in order to convert primary storage materials into usable compounds for energy liberation.

Proteins

The use of proteins and amino acids by fish as metabolic substrates has been well reviewed over the past few years by a number of authors (see Love, 1970; Woodhead, 1975; Driedzic and Hochachka, 1978; Blem, 1980; Cho *et al.*, 1982). Mammals appear to preferentially mobilize lipids from stored body reserves and initially do not use proteins when metabolic demands are greater than energy acquisition. Migratory fish on the other hand mobilize proteins early when relying on body reserves, especially during stages of food deprivation which occurs in a number of species (see Creach and Serfaty, 1974; Moon, 1983). Templeman and Andrews (1956) indicated an initial preferential use of proteins for migration in work with the American plaice, *Hippoglossoides platessoides*. The mobility of proteins also appeared to be higher initially in the North Sea cod, *Gadus morhua*, as well (Love, 1958). Deprivation of food leads to loss of protein before lipids or glycogen in the goldfish, *Carassius auratus* (Stimpson, 1965) and eels *Anguilla* sp. (Butler, 1968). Herring, *Clupea harengus*, cease feeding during their spawning migration when gonadal recrudescence is taking place. At this time collagen is deposited that is later used as an energy source before the herring can resume feeding after spawning (McBride *et al.*, 1960; Hughes, 1963; Thompson and Farragut, 1965).

There are a number of studies indicating that adult migratory Pacific salmon deplete their body reserves, a large proportion of which are in the form of muscle proteins. Unfortunately, many of these studies measured organic compounds in tissues before and at the end of the migration period and it is thus impossible to differentiate what was the sequence of use of the different stored compounds. More than half a century ago, Greene (1919, 1926) recorded a loss of more than half of the muscle tissue wet-weight of spring salmon, *Oncorhynchus tschawytscha*, during migration, which was equivalent to almost a third of the actual muscle protein. Duncan and Tarr (1958) and Idler and Clemens (1959) also showed similar protein losses in adult migratory salmon and indicated that female fish lost half as much again as migrating males.

Since fish appear to depend significantly on protein mobilization from muscle tissue for migration, it is of interest to ascertain where the proteins are mobilized from and the nature of the compounds released. Fish possess two types of muscle tissue, the so-called red and white fibres distinguishable by both morphological and biochemical criteria. White fibres constitute the largest percentage of the muscle mass and function largely anaerobically, whereas red fibres function mainly aerobically (see Driedzic and Hochachka, 1978). During slow swimming the red fibres appear to be active but during active locomotion, as would be the case in long arduous fish migrations, the white fibres become more active (Bone, 1966; Hudson, 1973). Under conditions of starvation, muscle proteins are preferentially used by the plaice, *Pleuronectes platessa*, and the white fibres lose their constituents before the red fibres (Johnston and Goldspink, 1973). Kutty (1972) has given some insight from his experiments with the Tilapia, *Sarotherodon mossambica*, after exercise. The fish were only starved for a short period (36 hrs) and, following exercise, NH_4^+ excretion levels were at such elevated values that they were indicative of protein catabolism providing all the required aerobic energy. As pointed out by Driedzic and Hochachka (1978) fish may excrete NH_4^+ directly as a product of protein catabolism and thus save energy by not having to produce excretory products such as urea or uric acid. They also point out that the anabolism of proteins from amino acids is energetically less expensive than the conversion of amino acids into glycogen or triglycerides. They also make an interesting suggestion that fish might be capable of producing unique storage proteins that can be utilized in times of energy demand. Mommsen *et al.* (1980) have provided some convincing evidence that adult migratory sockeye salmon display enzyme activities that are involved with the breakdown and utilization of proteins from white muscle tissue.

Although protein is lost from fish muscle during migration, the killifish, *Fundulus heteroclitus*, does not exhibit a decrease in protein synthesis during exercise (Jackim and LaRoche, 1973). It thus appears that the control over protein utilization during exercise is at the level of protein catabolism and not an inhibition of anabolism. The physiological controls over metabolyte mobilization are discussed in the next chapter. The enzymes involved with the breakdown of proteins in fish muscle are little understood but Driedzic and Hochachka (1978) provide a good review of the use of individual amino acids by fish muscle. Half the pool of free amino acids in fish muscle is comprised of glycine and histidine and both are major sources utilized during fasting. Demael-Suard *et al.* (1974) provided good evidence that in the tench, *Tinca vulgaris*, muscle glycine is converted to glucose in the liver following labelled glycine injections. Histidine, however, appears to be catabolized directly in the muscle tissue itself. Histidase, the enzyme responsible for the initial step in the breakdown of histidine has been found in the muscle tissue of mackerel, *Scomber japonicus* (Sakaguchi *et al.*, 1970). This enzyme has a much higher activity in fish muscle than in liver tissue. Wood *et al.* (1960) have measured histidine in tissues of migratory sockeye salmon, *Oncorhynchus nerka*, noting that the changes in levels indicate utilization mainly by muscle tissue. Other amino acids may also be broken down directly in muscle tissue without prior conversion to glucose as appropriate enzymes are found present in muscle tissue and in the carp, *Cyprinus carpio*, alanine and glutamic acid are oxidized faster than glucose (Nagai and Ikeda, 1973). Driedzic (1975) has indicated that carp white muscle is capable of generating NH_4^+ when at rest and levels increase when the fish is stimulated to swim. Thus, there is an indication that amino acids are oxidized during exercise as well as when maintaining a basic metabolic rate. The differences in red and white muscle fibres may be based in part on the types and ways in which the different amino acids are deaminated and enter the citric acid cycle for the eventual liberation of energy later on in the electron transport chain. These differences may thus be ways in which migrating fish are able to meet their bioenergetic demands.

Carbohydrates

Liver and muscle glycogen constitute stores of an energy-rich compound that can readily be mobilized and used for locomotion. Glycogen stores are particularly low in fish when compared to lipids and proteins and are not likely therefore to be relied upon for long distance migrations. During short periods of exercise, however, glycogen stores are probably very important (see Woodhead,

1975). Fish such as cod (Kamra, 1966), rainbow trout (Black and Tredwell, 1967) and kokanee salmon, *Oncorhynchus nerka* (McKeown *et al.*, 1975) all lose glycogen stores during starvation. In the case of the cod, the total body reserves in glycogen were not high enough to even maintain the basic metabolic rate of these fish, clearly indicating a need for other energy reserves. The low levels of carbohydrates in body storage deposits and only minor changes during conditions such as starvation may, however, be only a reflection of the low carbohydrate content in the diet of fish.

During arduous spawning migrations energy demands become severe, yet these fish maintain stored levels of carbohydrates. Atlantic and Pacific salmon maintain high levels of blood glucose during upstream migration (Jonas and MacLeod, 1960). Liver glycogen levels also remain high in adult migratory Atlantic salmon, *Salmo salar* (Fontaine and Hatey, 1953) and even increase significantly in anadromous sockeye salmon (Chang and Idler, 1960). Muscle glycogen in the migratory eel, *Anguilla rostrata,* is also maintained at high levels (Butler, 1968). It is interesting to note that starvation of kokanee salmon for 30 days reduced liver glycogen stores and tended to reduce plasma glucose levels, whereas exercise, continuously enforced for 24 hours had no effect on these carbohydrate parameters (McKeown *et al.*, 1975). Although some of the studies mentioned above on migrating salmon have measured carbohydrate levels in fish that have been actively swimming as well as not feeding, no reductions were observed. Thus there appears to be a physiologically different response to exercise than to deprivation of food with respect to utilization of carbohydrate stores. The effects of exercise override those of starvation during periods of migration. There also appear to be differences between young and mature fish such that older migratory fish maintain higher levels of carbohydrate metabolism (see Woodhead, 1975). Robertson (1955) also observed higher glucose levels in migrating steelhead trout, *Salmo gairdneri*, as compared to non-migratory fish. Since carbohyrdate stores are maintained or even sometimes increased during active migration (which is also at a time when the fish is often not feeding), there must be a utilization or even conversion of other substances. The breakdown and conversion of lipids and/or proteins (gluconeogenesis) is probably the source of these new carbohydrates.

Lipids

Lipids are a major source of energy for fish and are readily available in fish diets. Storage sites for lipids in fish are more extensive than in other higher vertebrates. Mammals store fats mainly in adipose tissue while in fish certain proportions of lipids are also stored in adipose-like tissue in viscera in such sites as along mesenteries and around the pyloric caecae (Farkas, 1967). Fats are also extensively stored in muscle and liver tissue in fish (Bilinski, 1969). Within the muscle tissue there is a differential storage between the red and white fibres, the former containing twice the levels (see Driedzic and Hochachka, 1978). The red muscle fibres also store fats intra- and extracellularly (George, 1962) and when in the cytosol they are surrounded by mitochondria (Lin *et al*., 1974). Lipids are also stored beneath the skin in some fish species. With such an array of storage sites available in some fish (for example, herring) lipids may account for as much as one third of the body weight.

Under experimental conditions, fish appear to mobilize lipids when they are either forced to exercise or are deprived of food. When kokanee salmon are starved for 30 days, circulating levels of free fatty acids drop, yet when forced to swim they tend to lose both stored muscle lipids and have reduced circulating levels (McKeown *et al*., 1975). Forced starvation of eels, *Anguilla anguilla*, for up to almost two years (Vladykov, 1964) or even three years (Bëotius and Bëotius, 1967) led to drastic reductions in lipid stores.

During natural conditions of starvation and active swimming that occur in many fish at a time of spawning migrations, lipid stores are mobilized and used as a major source of energy. Early work by Greene (1926) indicated that spring salmon started the freshwater phase of their anadromous migration with a muscle fat content of 15-20 per cent. This was reduced to 1-2 per cent by the time the fish arrived on the spawning grounds. Subsequent work by Idler and Bitners (1958) indicated that the sockeye salmon also lost significant quantities of lipid from the muscle tissue during upstream migrations. The sockeye also lost lipid at a greater rate from fat deposits other than in muscle stores. Female sockeye salmon utilized more lipids from the muscle than did males, but the female fish lost less from other fat deposits.

Gonadal recrudescence could account for a difference between the sexes as the ovary requires far more lipids during vitellogenesis than is needed for testicular maturation. Oceanodromous fish may not require as much energy for osmo(iono) regulation due to salinity transfers and do not have the necessity for locomotion against river or stream currents. However, some species travel long distances and still use lipids for an energy source. Herring store high levels of lipids during premigratory fattening and often mobilize most of these reserves during migration with a high proportion (15 per cent) going towards gonadal development (see Woodhead, 1975). Similarly, capelin, *Mallotus villosus,* also store large quantities of lipids that are almost entirely lost during migration (Winters, 1970). Atlantic cod mobilize stored lipids during winter months when food is relatively scarce and during migration lose considerably more body fats (see Woodhead, 1975). As with migratory salmon the female cod mobilizes more lipids and deposits them as ovarian vitellins.

Catabolism of stored lipids to free fatty acids provides migratory fish with a major fuel source for aerobic muscle metabolism. Krueger *et al*. (1968) have shown that the loss of lipid stores in the coho salmon, *Oncorhynchus kisutch*, depends on the distance travelled and thus the amount of muscular activity. Since red muscle fibres are capable of aerobic respiration, they have high concentrations of lipids and possess highly developed morphological characteristics such as numerous mitochondria. This muscle tissue type is probably responsible for muscle activity requirements for long distance migration. Red muscle fibres in a number of fish species have been demonstrated to exhibit capabilities of oxidizing free fatty acids at a much higher rate than white muscle fibres (see Driedzic and Hochachka, 1978). Red muscle fibres appear to be more prevalent in migratory species indicative of a role for increased muscular activity. Love *et al.* (1974) have shown that there are biochemical differences in the red muscle fibres of cod depending on whether they are from a migratory or non-migratory race. Walker (1971) has also shown that the red muscle fibres increase in diameter following forced swimming of the coalfish, *Gadus virens.*

Energetic Costs of Migration

No energetic value was assigned in the above section to each of the three major groups of organic compounds used by fish bioenergetically. Lipids, proteins and carbohydrates all have different energies available to fish once they have been assimilated or released from storage areas. However, the presence of these compounds in the diet of fish may not necessarily indicate the final energy that is available to that fish. Depending on the species, various organic compounds in the diet will exhibit different digestibilities and display varying amounts of excretory energy loss. At this point the fish has obtained a metabolizable energy source. Before these compounds can eventually be used for metabolism and growth they often require further modification before energy can be extracted. The deamination of amino acids is a good example. This process is a further energy loss to the fish as the so-called heat increment leaving the fish with a total usable net energy. Brett and Groves (1979) have reviewed the physiological energetics (see Figure 4.1) of the various organic compounds as they pertain to fish. They estimate that the energy available to fish from synthesized body carbohydrate, fat and protein is approximately 4.10, 9.45 and 4.80 kcal/g, respectively. The amount of energy available from these compounds in the diet of fish, however, is somewhat different because of the varying degrees of digestibility. They give estimates for the dietary compounds for salmonids in fresh water to be 1.6, 8.0 and 4.2 kcal/g for carbohydrate, fat and protein, respectively. From these approximations it becomes evident that fish do not digest carbohydrates well, but can more effectively use fats and proteins. These values vary depending on the species of fish and their ability to digest the different dietary organic compounds. For example, some herbivorous fish may display a metabolizable energy of dietary carbohydrate as high as 3.3 kcal/g.

Once a fish has obtained a net available energy supply the use to which these reserves may be put vary. Out of necessity a certain percentage of energy reserves must be allocated to metabolic demands. These energetic costs may simply represent the required energy to maintain a basic standard metabolism. In order for fish to achieve specialized activities such as swimming, feeding, aggression, spawning, migration, etc. energy must be available over and above that required for basic metabolic purposes. In order to liberate energy from

Figure 4.1: Average partitioning of dietary energy for a fish for every 100 calories. Non-fecal energy is mostly that excreted as ammonia and urea. The numbers in brackets indicate a range which will depend on parameters such as activity level, temperature, fish, size, age and ration size. (I), rate of ingestion; (E), rate of excretion; (M), rate of metabolism; (G), rate of growth; (M_F), feeding metabolism; (M_A), active metabolism; (M_S), standard metabolism.

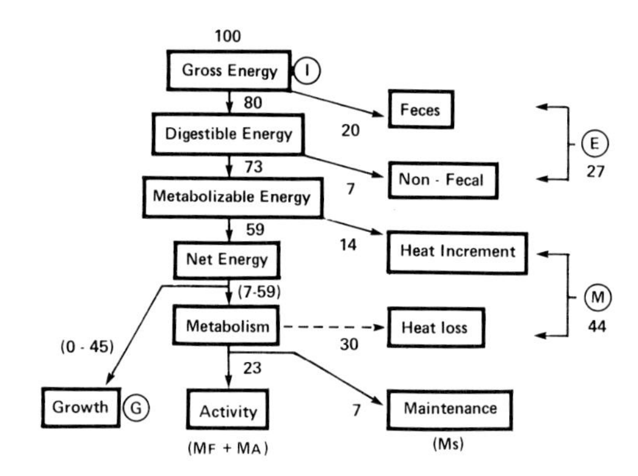

Source: redrawn from Brett and Groves, 1979

organic compounds a certain amount of additional energy must be potentially available since fish lose a proportion of energy through excretory or digestive processes. As compounds are converted to usable forms to enter the Embden-Meyerhoff pathway or the citric acid cycle to yield energy, there will be a proportion of energy lost due to the required conformational changes as well as compounds that become unusable for one reason or another. Excretory products often require energy as well in order for them to be converted to storage, transport or elimination forms. Digestive and assimilation processes are not always maximally efficient and some of these specific functions are not even available to some fish species, which leads to a further energy loss.

When all the energetic demands of metabolism are met as well as compensations for energy losses, there may still be an excess amount of energy-rich compounds still available. At this point there are a number of options available. The fish now would be capable of growth but in different ways. Although an individual fish may survive for a finite time when provided with enough energy to meet metabolic demands, the survival of the population of fish to which the individual belongs would not continue unless extra energy is available for growth. In this respect growth would be required for specific tissues such as during gonadal recrudescence. Growth may also occur in other specific tissues such as muscle and bone and give the individual greater advantages for inter-specific and intraspecific behavioral interactions which ultimately may also lead to reproductive success. Additional energy reserves may also form part of growth by being deposited in various storage sites which can be called upon later for a metabolic energy supply or conversion to differential growth such as in gonadal tissue.

Metabolic Costs

During any phase of the life cycle of all fish species a certain amount of energy must always be available to meet maintenance requirements. From experiments with warm-blooded vertebrates the minimal energy requirements have been termed basal metabolism. Since experiments were difficult to perform with fish, this minimal energy output was termed standard metabolism (Krogh, 1914). Normal fluctuations in fish activity patterns were termed routine metabolism (Fry, 1957; Beamish, 1964) whereas maximum energy expenditure is termed active metabolism. New equipment and the resulting data which have now become available indicate that basal and standard metabolism are synonymous. Brett and Groves (1979) have compared the standard metabolic rates between species and discuss the differences exhibited among polar, temperate and tropical species. Species living in polar regions and, to a lesser degree, those from tropical regions, exhibit metabolic compensation to live at the low and high temperatures. The standard metabolic rate of these fish is higher than fish from temperate climates. The minimal metabolic rate observed for fish is approximately 20 mg of O^2/kg/hr, well below that recorded for homeothermic vertebrates. Within a given species the standard metabolic rate is also affected by temperature. This was first described by Ege and Krogh (1914) for the goldfish and is called Krogh's standard curve. The rate of increase of the standard metabolic rate with an increase in temperature is well illustrated and discussed by Fry (1971). The standard metabolic rate also depends on the size of the fish. Larger fish of the same species or between different species generally display a reduced rate per unit weight.

The vast majority of energy requirements for fish come about as a result of increased activity related to such behaviors as food acquisition, predator avoidance and migration. Fish exhibit bursts of activity but, if at too high a level, this will result in exhaustion since the cardiovascular system cannot supply enough oxygen. However, fish can elevate their standard metabolic rate up to a maximum depending on oxygen consumption and at this level may maintain this increased active metabolic rate for prolonged periods of time. As with the standard metabolic rate, the active metabolic rate also increases with temperature but only up to a maximum where often it declines after a further temperature increase (see Fry, 1971). This maximum point varies with different fish species depending on their thermal ranges. Jones (1971) hypothesized that higher temperature caused a decrease in active metabolism because of limitations brought about by the capacities of different species to acquire and deliver oxygen to active tissues. Also, the oxygen content of water decreases with an increase in temperature (Fry, 1971). Nevertheless, active fish species are still capable of raising their metabolic rate an order of magnitude above standard metabolic rates and can do so for prolonged periods. Furthermore, some fish can also switch to anaerobic metabolic pathways and increase their active metabolic rates even many times further. During periods of extreme activity when anaerobic energy sources are used, the fish must subsequently be able to pay back the oxygen debt and these extremely elevated metabolic rates are thus only possible for limited periods of time.

Brett and Glass (1973) have studied the effect of size on active metabolic rates of sockeye salmon and found only a minor or negligible effect. The decrease in respiration rate of muscle tissue in a larger fish is probably offset by the fact that larger fish have proportionately more muscle tissue than smaller fish of the same species. When swimming performance is compared to length of the fish instead of weight these differences in size become apparent (Beamish, 1978). Generally, longer fish appear to do less well when compared to shorter fish of the same species, at least when tested at burst or maximum speeds. The longer fish would appear to have proportionately more muscle and possibly, therefore, relatively more carbohydrate or oxygen reserves but encounter a problem of increased hydrodynamic drag which apparently outweighs the former advantages. The importance of weight to length relationships (or the so-called condition factor) become important when considering energy expended during active metabolism. Green (1964) has conducted interesting experiments with different stocks of brook trout, *Salvelinus fontinalis*, to ascertain the effects of condition factor on swimming performance. Fish from wild stocks, but of equal length as domestic stock fish, performed better although their weight was less. The higher fat content of the domestic stock and thus their higher condition factor, may have been the reason for their poorer performance. The fish with a higher condition factor may have to supply its tissues with more oxygen but may, however, only have similar morphological capacities for acquiring and transporting required metabolites and gases as does the fish of lesser weight. An increase in the condition factor may also increase hydrodynamic drag as well. It is interesting to note that young salmonids exhibit a decrease in their condition factor during smoltification which occurs just prior to the time of seaward migration. Fish of the same size (length and/or weight) may also display different energetic requirements depending on their sex and degree of infestation by parasites. Beamish (1978) has reviewed the literature with regard to these last effects and documented the studies that indicate the particular parasitic species and the degree to which swimming performance is affected.

The maximum sustained active metabolic rate is largely due to the amount of oxygen that can be delivered to the tissue which in turn depends on the morphological and physiological characteristics of the species such as found in ventilation mechanisms, gill characteristics, and cardiovascular adaptations. These characteristics of any given species depend on an adequate oxygen supply from the environment and if the available ambient oxygen concentration changes then the active metabolic rate will be affected. If oxygen concentrations become too low fish will be unable to maintain sustained or prolonged speeds. Kutty (1968) and Kutty and Saunders (1973) termed this point as the critical oxygen concentration. Beamish (1978) has reviewed a number of studies that indicate that there are a variety of fish species studied in which lowered ambient oxygen concentrations reduce sustained and prolonged swimming performance. Bursts of swimming activity are unaffected by environmental oxygen concentrations since this type of activity depends on anaerobic processes, although the recovery period between bursts of activity would be increased. Beamish (1978) also pointed out that carbon dioxide levels would also be likely to affect swimming performance since the binding affinity of haemoglobin for oxygen and the metabolic rate of fish are affected by this metabolyte. Little work has been done in this area except that of Dahlberg *et al.* (1968), who described a decline in swimming performance with increasing carbon dioxide levels in coho salmon. They failed to observe any effects of carbon dioxide on swimming performance of largemouth bass, *Micropterus salmoides*.

Fish responding to ambient salinity concentrations other than isosmotic environments, must expend energy to actively transport ions into the environment when they are in a hyperosmotic medium and sequester salts from their environment when exposed to hypo-osmotic surroundings. As the ionic concentration gradients increase the number of ions to be transported would also increase and therefore there would also be an increased energy demand. Glova and McInerney (1977) found only slight effects of different salinity gradients on the swimming performance of coho salmon fry and smolts. Swimming performance in rainbow trout (Rao, 1968) and *Tilapia nilotica* (Farmer and Beamish, 1969) appeared to be independent of salinity but the metabolic expenditure did vary, which is indicative of an increased energy demand for ionic regulation.

The active metabolic rate of fish is affected by a variety of environmental contaminants. These effects may be direct or indirect depending upon the mode of action of the particular pollutant or the fish species in question. Many environmental toxicants have effects on specific fish tissues and as such may cause an expenditure of energy that would diminish that available for other functions during increased periods of activity. For example, increased zinc concentrations in the water are known to decrease circulating levels of insulin in rainbow trout (Wagner and McKeown, 1982). Under the same conditions liver glycogen levels are also decreased and plasma glucose elevated. Thus, if this particular toxicant interferes with one of the major regulatory factors for carbohydrate metabolism, sustained metabolic rates would also be affected. Similarly, reduced environmental pH that occurs during such times as acid precipitation, increases energy demands due to adverse affects of increased hydrogen ion concentrations. Reduced environmental pH is known to cause an increase in rainbow trout gill epithelial permeabilities (Parker and McKeown, unpublished). Under such conditions the trout lose ions in a freshwater environment but try to compensate by increasing specific gill ATPase enzyme activities (Watson *et al.*, 1984). This increased energy demand would also affect active metabolic rates for other activities such as swimming. Many different fish react to a wide variety of environmental toxicants by secreting mucous over the gill epithelium, presumably in an attempt to restrict the influx of the pollutant. At the same time the diffusion distance for oxygen is increased and if exposure is continued the fish may die of anoxia. At intermediate levels, decreased oxygen uptake would decrease metabolic rates. Other toxicants such as the organophosphorous insecticide acephate seem to affect respiration rates of rainbow trout directly (Geen and McKeown, unpublished). Although the mechanism is unknown, relatively low levels of this pesticide increase oxgyen uptake and presumably utilization. This may only be a reflection of an increased metabolic demand because of some adverse tissue effect or an increased demand for some defense mechanism, but nevertheless is indicative of enhanced energy demands.

These extra energy demands may seem superfluous when considering the overall bioenergetic costs of a fish migration but for some species that push their energetic demands to the extreme, the additional costs may be overwhelming. Unfortunately, many industrial effluents are discharged into waters at or near the entry of rivers into the ocean. At these particular points in the migration route of spawning anadromous fish the additional energy demands occur at a time when the fish is changing salinity environments and is about to undertake the arduous upstream portion of its migration. Not only might these pollutants give migratory fish problems of olfactory orientation but they might also cause devastating effects bioenergetically.

Costs of Feeding

Active swimming is the major behavior during migration that utilizes the energy reserves of a fish. Feeding activity and metabolism, however, also require substantial amounts of energy. Even though some fish species may have the opportunity to feed during migration, the energy input may be outweighed by the energy required to obtain the extra energy source. Brett and Groves (1979) outline a number of examples to illustrate the magnitude of the energy required for feeding metabolism (Figure 4.2). When Atlantic cod were fed at high ration their metabolic rate rose 4.7 times the basal rate (Edwards *et al*., 1972). Likewise, the reef fish, *Kuhlia sandvicensis*, increases its routine metabolic rate of 76 mg O_2/kg/hr to 180 mg O_2/kg/hr shortly after the feeding of a single ration (Muir and Niimi, 1972). The circadian feeding pattern of sockeye salmon showed an increase during feeding from 170 to 370 mg O_2/kg/hr when the fish were swimming slowly (Brett and Zala, 1975). The feeding metabolic rate thus varies between species but as summarized by Brett and Groves, may increase as much as 5.8 times the standard metabolic rate for some fish. The authors also indicate that the feeding metabolic rate varies within a given species depending on different factors. The feeding metabolism rate is thus highly ration-dependent, the higher rations demanding a higher feeding metabolic rate. Some species exhibit a linear response to size of ration up to a feeding maximum whereas other fish species display a sigmoidal relationship. An increase in temperature likewise causes an elevation in feeding metabolic rate. The expenditure of energy associated with feeding may be due to a number of different factors.

Figure 4.2: The feeding cost of carp. The change in metabolic rate with increasing levels of daily ration. The feeding cost also changes with temperature.

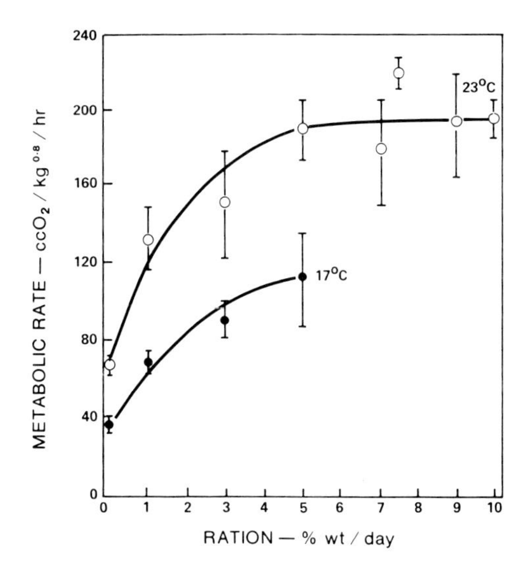

Source: redrawn from Brett and Groves, 1979, after Huisman, 1974

The acquisition and ingestion of food obviously requires various locomotory and alimentary canal tissue activities as well as nervous tissue excitability but there is also a requirement of energy to transform foodstuffs into a usable form. This additional energy for biochemical transformations has been termed heat increment or specific dynamic action. Brett and Groves (1979) outline a number of well-designed experiments with fish which indicate that the heat increment for some fish may be as high as one third of the energy of the ingested food. The heat increment within a given species appears to vary with the ration size, weight of the fish and type of organic compound ingested.

Migrating fish expend large amounts of energy for the required muscle activity for locomotion but also require an expenditure of energy for other activities in addition to feeding. Brett and Sutherland (1965) and Brett (1973) have conducted studies with the pumpkinseed sunfish, *Lepomis gibbosus*, and sockeye salmon, respectively, to ascertain the amount of energy required for aggressive behavior and in both cases measured large increases in energy expenditure associated with intraspecific contests. Aggressive behavior often occurs at the end of spawning migrations for defense of a territory and is thus a cause of an energy requirement that must be prepared for during a premigratory fattening period. Likewise, spawning behavior itself (nest building, spawning, courtship, parental care, etc.) requires, in many species, a considerable expenditure of energy although the detailed measurements have not been obtained in most instances.

Daily vertical migrations are interesting with respect to energy expenditure. There is an added energy requirement to move vertically through the water column and some appropriate measurements have been obtained (see Brett and Groves, 1979). However, the fish seem to have a net energy gain by retreating daily to waters of lower temperature where metabolic energy demands are less. Although the above activities require an expenditure of energy they do not reach the demands placed upon a fish such as during the time of an upstream spawning migration that occurs in fish like Pacific salmon.

Indirect Costs – Energy Loss

Energy available for metabolism and/or growth is preceded or accompanied by energy losses due to assimilation and excretory processes. Egested wastes not only contain undigestible food but also losses of endogenous compounds. The passage of foodstuffs through the digestive tract of fishes is accompanied by a loss of epithelial cells, mucus, metabolized enzymes, secreted hepatic wastes, and bacteria. Depending on the diet of individual fish species, considerable differences with respect to losses in fecal energy occur. Herbivorous fish can thus lose up to 50 per cent of the energy of ingested food whereas the loss for some omnivorous species feeding on various invertebrates can be at a lower loss level of 25-30 per cent. Piscivorous fish-eating species can exhibit even fewer losses of approximately 20 per cent. The fecal energy loss may also vary considerably as the fish matures and changes it food sources. Brett and Groves (1979) have outlined a number of studies that also indicate the involvement of various physical and biological factors that affect the degree of fecal energy loss. These parameters are such things as temperature, size of meal, frequency of feeding, weight of fish and pre-starvation. Both rainbow trout (Brocksen and Brugge, 1974) and brown trout, *Salmo trutta* (Elliott, 1976) display a reduced fecal caloric loss when fed at higher temperatures. The fecal loss of largemouth bass has been shown to be reduced when feeding was extended from once per day to every five days (Blackburn, 1968). Elliott (1976) also demonstrated that the brown trout had a relatively reduced fecal caloric level when fed smaller rations.

Depending on the species and the type of diet, a certain proportion of the energy content of ingested food is lost in the feces as a result of indigestible and non-assimilable materials. In addition, there is a further energy loss as energy rich compounds are metabolized as an energy source or used for growth. There is an energy loss from all types of assimilated organic compounds but particularly from proteins. Many absorbed lipids and carbohydrates are in a form that can be used readily for energy liberation but proteins require deamination and the necessity that the resulting nitrogen be excreted in the form of ammonia or urea. Fish can absorb a high proportion of protein from their diet and, unlike other vertebrates, use assimilated amino acids more readily as an energy source than they do for other organic compounds. Brett and Zala (1975) have made interesting calculations from studies on sockeye salmon fingerlings in order to estimate the amount of energy lost from conversions and excretion of nitrogenous wastes after assimilation of proteins from the diet. After taking into account the amount of energy lost from breakdown of amino acids from an endogenous source which was obtained from studies on starving fish, they calculated that 27 per cent of the energy was lost from the nitrogenous intake. Since many fish depend on protein sources to a large extent, this is a significant energy loss. As wtih energy losses due to indigestible materials, Elliott (1976) demonstrated that various factors could also affect the energy loss due to excretion of nitrogenous wastes. This latter loss was found to be greater with increasing temperature and decreasing level of intake.

Respiratory and Circulatory Costs

Increased metabolic rates require the delivery of metabolites to and removal of by-products from the active tissues involved. The acquisition and transport of respiratory gases to and from metabolically active tissues depends on the respiratory and cardiovascular systems which themselves place additional energetic demands on fish. An increase in the activity of fish is usually accomplished by an elevation in oxygen metabolism although bursts of swimming activity may be accomplished by anaerobic respiration of white muscle fibres. Jones and Randall (1978) have reviewed the involvement of the respiratory and circulatory systems with increases in the activity of fish as well as the relationships between oxygen metabolism during increased exercise. The difference between active and standard rates of oxygen uptake should give a close approximation of the amount of oxygen available to muscles for locomotion. However, the heart and respiratory muscles require more oxygen during conditions of increased exercise and the pumping of more water over the gills causes increased ion and water fluxes, which in turn causes energetic demands for osmo(iono)-regulation. The relationship between oxygen uptake and increased activity was investigated by Brett (1964) with sockeye salmon who concluded that oxygen uptake increased exponentially with swimming speed. Jones and Randall (1978) discuss other factors that influence oxygen uptake such as temperature and the size of the fish. The oxygen utilization of muscles is not equally efficient at all power outputs which would also contribute to a non-linear increase in oxygen uptake with increasing exercise. The maximum activity of fish and thus their upper limits of oxygen uptake and utilization are no doubt limited or terminated by fatigue. Whether fatigue is brought about by exhaustion or cellular energy supplies or some other factor(s) such as acidosis from anaerobic metabolism is still equivocal.

Whatever the relationship is for a given fish species between oxygen uptake and increased activity, the respiratory system can be altered to meet the increased demands. More oxygen is made available to fish during exercise by increasing the amount of water passing over the gills per unit time. Fish increase the amount of water flowing over the gills both by an increase in ventilation rate as well as an increase in the volume of water pumped during each ventilation cycle. As fish increase their swimming speed, they may still irrigate the gills with the same water volume as from ventilation pumps by simply opening the mouth. This type of ram ventilation in some fish may reduce drag during swimming and at the same time may save energy from branchial muscular activity that can be used by the swimming musculature (see Jones and Randall, 1978). More oxygen may become available to fish during exercise by passing more water over the respiratory surface but a similar oxygen availability may be realized by an increase in blood flow through the fish gill. Oxygen diffuses into the blood in the secondary lamellar vascular space of the fish gill. The channels in this space are formed by pillar cells and during times of increased oxygen demand these cells might change their shape and thus alter the blood flow distribution through each lamellae. Alternately, more blood may be delivered to the gill by increasing the number of individual lamellae. Distal lamellae have smooth muscles around afferent and efferent arterioles and during exercise, adrenergic stimulation may dilate vessels to ensure an increased blood supply to the gills (see Jones and Randall, 1978).

Another means to increase oxygen availability during exercise is by elevating transport of gases in the blood. There is an indication that oxygen is utilized more by the tissues of the carp, (Secondat, 1950) and rainbow trout (Kiceniuk and Jones, 1977) since the venous oxygen concentration is reduced following exercise. This reduced oxygen tension would also facilitate oxygen uptake in the gills since the oxygen concentration gradient between the water and blood would be greater. Erythrocytes are released from the spleen in exercised rainbow trout (Stevens, 1968) as well as an increased diuresis (Wood and Randall, 1973) which leads to an increased hematocrit. The increased hemoglobin concentration during exercise will thus enhance the oxygen carrying capacity of the blood and at the same time would decrease the cost of cardiac work. Carbon dioxide levels in the venous circulation increases during exercise and thus CO_2 transport at the gill will be enhanced by increased concentration gradients and by the action of gill epithelial carbonic anhydrase activity (see Jones and Randall, 1978).

During exercise fish may increase the amount of oxygen delivered to the tissues by changes in the cardiovascular system such as heart rate, stroke volume and/or oxygen concentration differences between anterial and venous blood. Experimental results from a number of studies involving a wide variety of fish species indicate a variable response in heart rate to increases in exercise (see Jones and Randall, 1978). Some of the variations in observed tachycardia following exercise may be related to different species being generally active or sluggish since more active fish such as salmonids display lower heart rate increases following exercise. In contrast to the variable heart rate responses of different fish species to exercise, different studies appear to agree that the stroke volume of the heart of fish increases during forced swimming. Oxygen uptake during exercise in fish may be enhanced by increases in blood pressure and reduction in the total peripheral resistance to blood flow. Arterial blood pressure is known to increase during exericse in teleost fish and consequently may cause a passive vasodilation to reduce resistance although neural and hormonal factors may also be involved (see Jones and Randall, 1978).

Energy Gain – Growth

This subject has been well reviewed (see Brett, 1979; Brett and Groves, 1979; Ricker, 1979) and only a short summary will be included here. By necessity, fish must utilize energy reserves for general maintenance as their first priority. Other activities such as feeding must also be attended to in order to meet the demands of standard metabolism. Extra energy must be made available for fish to grow and reproduce in order for one generation to follow another. Almost all fish can achieve growth if the available food or ration is increased. Some fish species continue to grow with increasing ration in an almost linear relationship up to a maximum growth rate or limit of food that can possibly be ingested. Other fish species grow rapidly at relatively low rations but at higher rations their efficiency of food utilization is less and growth rate slows. Growth rate relative to body size also varies within an individual fish. As long as the fish has learned to feed and has developed its digestive system, the smaller the fish, the faster its relative growth rate. As fish increase in size and age their growth rate declines, as does their metabolic rate. If a sufficiently high ration is available, smaller fish use more energy for growth than for metabolism. Older fish near their upper size limit have reduced growth and metabolic rates but growth rates decline more rapidly so that they are near zero when metabolic rates are now higher (Figure 4.3).

Figure 4.3: The relationship between the energy deposited in growth and that expended in total metabolism as affected by the size of growing sockeye salmon. As the fish grows the ratio of growth to metabolism becomes less than one which may be a reflection of a decrease in food conversion efficiency that occurs with increasing size.

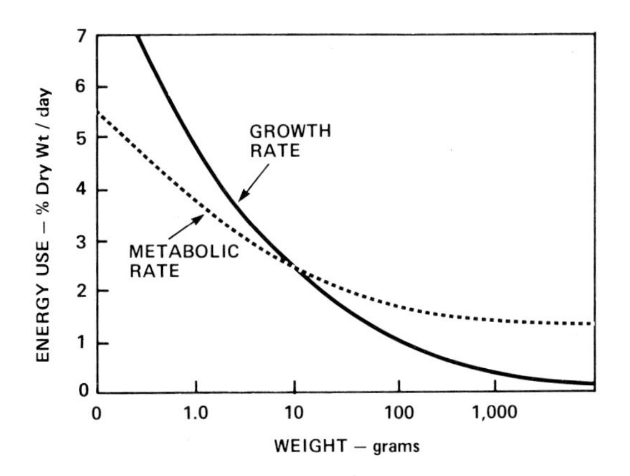

Source: redrawn from Brett and Groves, 1979, after Brett, 1970

Since the growth rate of fish decreases with age and size even though an adequate diet is provided, the conversion efficiency must therefore be less. This is true for the gross conversion efficiency which is the ratio of growth rate to ration expressed as a percentage. It is also true for the net conversion efficiency which takes into account the amount of a ration required for maintenance. Gross conversion efficiencies of maturing fish generally range from 10-25 per cent (see Brett and Groves, 1979), however, efficiencies up to 80 per cent have been observed in some fish when converting yolk. Some small fish also exhibit high efficiencies of over 50 per cent. Herbivorous fish generally have lower gross conversion efficiencies (10-20 per cent) but some species such as the grass carp, *Ctenopharyngodon idella*, may reach efficiencies over 40 per cent (Stanley, 1974).

Conversion of ingested food into new tissue requires the diet to contain high levels of protein ranging from 35-55 per cent (Cowey and Sargent, 1979). Given an appropriate diet, fish still exhibit differences in growth and conversion efficiencies depending on various environmental parameters. Temperature may control growth due to an effect on the rate of metabolic reactions. Depending on the ration and species of fish, varying temperatures within tolerable ranges show different effects on growth and conversion efficiencies (Averett, 1969; Brett *et al.*, 1969; Elliott, 1976). When fish are exposed to water that is not isosmotic with their blood they must expend energy for osmoregulation. A number of studies have indicated that when fish are exposed to salinities that impose a metabolic demand, their growth rates decrease (see Brett, 1979). Photoperiod also affects the growth of fish but this is probably due to an indirect effect of modifying endocrine cycles. Other factors such as oxygen affect growth rates, but again do so indirectly by limiting metabolic processes requiring energy derived from oxidative phosphorylation.

5 PHYSIOLOGY

Introduction

How can fish adapt physiologically to the demands of migration? This chapter will attempt to review some of the salient physiological features that fish display during migration. A fish functions as an integral unit and it is therefore perhaps artificial to discuss any one physiological aspect in isolation. However, this will be attempted in the interest of logical organization, but it should be borne in mind that no one physiological process operates without being affected by, or influencing others.

Given the diversity of fishes it is not surprising that the physiological adaptations themselves also differ between species. Many fish species migrate entirely in fresh or sea water whereas others migrate from one to the other. Not only are there vast differences in osmo(iono)regulation but also in other functions such as temperature adaptation, orientation mechanisms, energetics, photoperiodicity, reproduction and behavior. Some fish migrate over relatively short distances whereas others, such as eels or salmon, traverse distances of thousands of kilometers. Such long distance migrators must have specialized mechanisms for nutrient acquisition, storage and utilization.

Many physiological mechanisms are not only preceded and followed by other associated functions but may also overlap and be virtually inseparable from each other. For example it would be impossible to dissociate sexual behavior from other functions that take place during gonadal development such as metabolic changes or hormonal alterations. It is also difficult to distinguish exactly when a migrating fish will respond or adapt to a particular environmental set of conditions. A fish may thus display a preparative or an adaptive physiological response. To illustrate this point a young salmon preparing to migrate to sea would be a good example. While such a fish is in fresh water it goes through the process of smoltification and thereby becomes physiologically prepared or pre-adapted to regulate its body fluids in sea water even before it encounters such an increased salinity environment. The process of actually adapting to the increased osmotic pressure of sea water of course only starts when the smolt enters the ocean.

Another complication for studying the physiological basis for fish migration is due to the fact that a number of different functional mechanisms are occurring simultaneously even though they appear to be related. Many adult salmon cease feeding while undergoing long arduous migrations into fresh water. During this time these fish require energy for swimming as well as for gonadal maturation. This necessitates the mobilization of energy reserves for catabolic processes, on the one hand to liberate energy from stored metabolites and anabolic processes, on the other hand to transfer and change organic compounds into gonadal tissue and their associated products. These require similar compounds to be liberated but for vastly different purposes.

This chapter will discuss physiological processes which are central to fish migration. Osmo(iono)regulation is essential for all fish in aquatic or marine environments. Both potamodromous (fresh water) and oceanodromous (sea water) migrants must therefore regulate water and ion levels in their bodies and have adaptational abilities since water and ion movements will change as the fish expend energy and are exposed to varying temperature regimes. Those migrating fish species which are diadromous require refined mechanisms to adjust to drastic salinity changes. Migrating eels and salmon are commercially important species and display long catadromous and anadromous migrations respectively for which there is now considerable osmo(iono)regulatory information available. The process of smoltification will also be discussed since it is involved with the acclimitization of young freshwater salmonid fishes to sea water and is thus closely related to osmotic and ionic regulation. The parr-to-smolt transformation not only involves physiological changes for regulation in sea water but also encompasses morphological (coloration, skin mucus, body shape, scale modifications, and development of specialized gill and endocrine cells) and metabolic (body fat content, serum proteins, guanine deposition, visual pigments, glycogen stores, liver vitamins and body protein composition) alterations. Migration is an energetic cost to fish and thus physiological controls for metabolite deposition before, and mobilization of reserves during, migration will be considered. Many fish migrations are, as we have seen, for reproductive purposes and are therefore movements towards spawning sites. Gonadal maturation is often taking place during this time with associated structural and physiological changes. During any phase of a migration the fish must be capable of orienting to various environmental cues in order to find its predetermined destination. To understand orientation mechanisms, studies on sensory physiology are important. In order to migrate successfully a fish needs to develop appropriate behavior patterns at the desirable times. The physiological basis for behaviors such as migration initiation, feeding and swimming activity is now being investigated and is increasing our knowledge and appreciation of how fish can perform the many and integrated behavioral patterns constituting an entire migration.

Osmo(iono)regulation

Living teleosts occur from freshwater to hypersaline environments. Many species are restricted to a narrow range of salinities (stenohaline) whereas others are capable of regulating within a wide range of osmotic conditions (euryhaline) and may therefore migrate through them. Even though some species are restricted to a particular salinity environment there is a wide range of strategies between species for conforming or regulating water and ionic balance. Before discussing how various fish adjust their physiology to changing salinities during migration, this section will outline how fish in general adapt to one particular environment. Physiological adaptions to different salinity environments may involve behavioral manifestations but the basic controlling factors are elaborated by the endocrine system.

The blood of fish in fresh water is hypertonic to the environment which results in water and ion fluxes in opposite directions. Water is continually passing across permeable membranes into the fish by osmosis due to the concentration gradient. At the same time ions are diffusing across membranes and being lost to the environment. These fish solve the osmotic influx problem by producing a copious dilute urine. Lost ions are sequestered from the aquatic environment by active transport mechanisms mainly at the level of the gill epithelia (Figure 5.1).

Figure 5.1: Movement of water and ions in freshwater and seawater fish. In fresh water the fish gains water by osmosis and loses ions to the environment. These fish eliminate the excess water by producing a copious hypotonic urine and regain lost ions by the use of food salts and ion transport mechanisms across the gill epithelia. In sea water the fish loses water by osmosis and gains ions from the environment. These fish regain the lost water by drinking sea water. The transport of this ingested water across the intestinal wall is accompanied by ions as well. These fish eliminate the excess ions by transport across the gill epithelium.

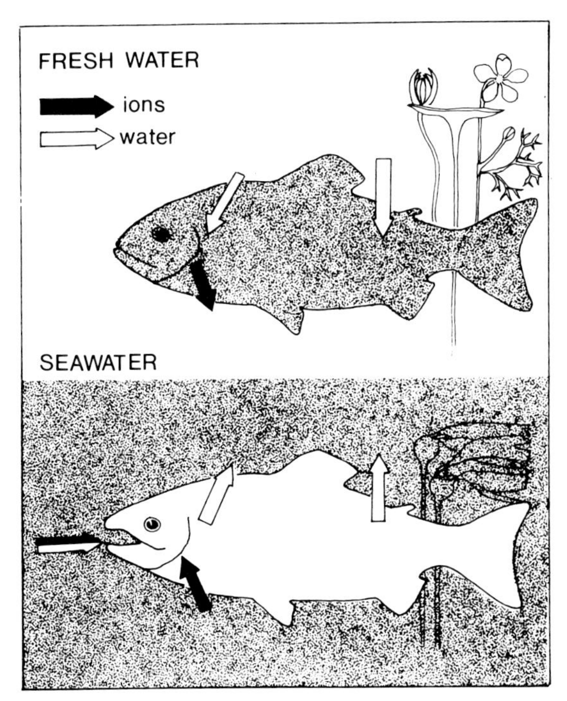

Marine fish are hyposmotic to their environment and consequently these fish have opposite osmotic and ionic problems than those experienced by freshwater species (Figure 5.1). Marine fish thus lose water by osmosis and gain ions down the concentration gradient. These fish regain the lost water by drinking sea water. This behavioral response adds, however, to the problem of elevated ion concentrations that also accumulate via diffusion. These excess ions are actively transported across membranes back to the marine environment. The gill epithelium is a major site for such transport and to a lesser degree the kidney is involved. However, in order to produce urine, even if it is hypertonic, a certain loss of water will also occur. Many marine fish species thus produce very small amounts of urine. Other species conserve water by having vascular shunts away from the kidney to reduce the glomerular filtration rate. Urine output is reduced in yet other species by a reduction in the size of the glomerulus that can even be reduced to an aglomerular kidney in some species.

Osmotic and ionic balance is achieved differently for other marine fish such as chondrichthians and myxinoid cyclostomes. Most sharks and rays solve their osmotic problem by retaining urea and trimethylamine oxide in their blood so that they are slightly hyperosmotic to their environment. There will thus be an influx of water by osmosis, but ions will also enter by diffusion mainly across the gill surfaces since their internal ion concentrations are lower than in sea water. Ions are eliminated by active transport across the gill epithelium and to a lesser degree salts are also lost via the urine and secretions from the rectal or salt gland. Hagfish maintain their blood osmotic pressure slightly above that of sea water by retaining high ionic concentrations. Osmotic problems are thus solved but certain ions still tend to diffuse into the blood. Excess ions are removed by hagfish via the kidney and skin mucous secretions. Intracellular fluids, however, have lower ionic concentrations but the osmotic concentration is still maintained similar to that of the blood due to the presence of trimethylamine oxide. In this case ions are thus regulated to a greater extent between the blood and other tissues than between the blood and the external environment.

Fish gills and kidneys are the major sites for water and ion exchanges but other tissues such as skin, bladder and intestine are also involved. The gill epithelium represents a relatively large surface area of the fish for purposes of gas exchange. At the same time this large surface area accounts for the major fluxes for both water and ions. Although water passes across fish gill epithelia passively, branchial membrane water permeabilities can be altered to affect the degree of osmotic water flow. Depending on the concentration gradients, ions are actively transported by specialized chloride secretory cells (Keys-Wilmer cells) located at the base of the gill lamellae (Figure 5.2). These cells have the ability to transport ions through adenosine triphosphatase enzyme (ATPase) mediated exchanges which employ ATP as the energy source. Na^+ and K^+ levels in the blood thus appear to be regulated by a carrier mediated exchange of Na^+ for K^+ in the form of a Na^+-K^+-activated ATPase (see Evans *et al.*, 1982). There is evidence that Na^+ may be exchanged for other ions besides K^+. Respiratory exchange of CO_2

Figure 5.2: Diagrammatic representation of a fish gill, gill leaflet and chloride secretory cell. (A), chloride secretory cell; (B), respiratory epithelium; (C), blood vessel; (D), tight junction; (E), epithelial cell; (F), apical cavity; (G), glycogen granules; (H), Golgi apparatus; (I), mitochondria; (J), nucleus; (K), tubular system; (L), basal lamina.

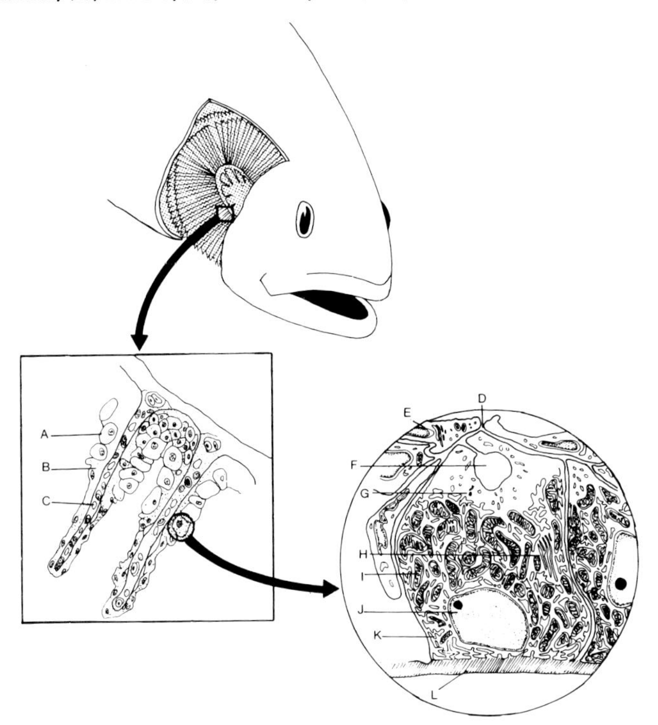

with the external environment is facilitated by hydration to HCO_3^- and H^+ by the enzyme carbonic anhydrase (Bundie, 1977). The resulting H^+ can then be exchanged for Na^+ by a Na^+-H^+-ATPase (Maetz, 1973; Payan and Maetz, 1973). Alternatively, the H^+ can be eliminated via the gill in combination with nitrogenous wastes in the form of ammonium. The excretion of ammonium is also thought to be an active exchange process mediated by a Na^+-NH_4^+-ATPase (Butler and Carmichael, 1972; Payan and Maetz, 1973). Bicarbonate ions which are the remaining by-product of CO_2 hydration are thought to be eliminated via the gill, mediated by a HCO_3^--Cl^--ATPase (McCarthy and Houston, 1977). There is evidence suggesting that the divalent cations such as Ca^{++} and Mg^{++} are also transported across fish gill epithelia via active transport mechanisms mediated by Ca^{++}-ATPase (Ma *et al.*, 1974; Fenwick, 1979; Ho and Chan, 1980) and Mg^{++}-ATPase (Doneen, 1981) respectively. The details of ionic exchanges across gill epithelia of fish in fresh and sea water have been well reviewed recently by Evans *et al.* (1982).

The teleost kidney is not capable of producing a hypertonic urine as is the case for higher vertebrates. Thus, urine production in fish exposed to a marine environment would only enhance water loss problems that the fish already has because of osmotic water efflux. The production of a copious dilute urine for fish in fresh water, however, is a useful means by which the excess water can be eliminated. Such marine species of fish can nevertheless regulate water loss during nephric excretory processes by mechanisms involving the urinary bladder. Some euryhaline fish exposed to sea water or some purely marine forms can reabsorb water and/or ions across the bladder epithelium to conserve water in their dessicating environment. Certain freshwater species also use the urinary bladder for ion regulation. These fish need to sequester ions from their hypotonic environment via gill epithelia and reduction in ion losses by reabsorption across the urinary bladder epithelium helps conserve otherwise lost ions (see Bentley, 1971).

The intestinal epithelium, in addition to the gill and kidney, also has a role in water and ion regulation. The intestine of fish adapted to fresh water contains a Na^+-K^+-APTase for actively transporting Na^+ to the blood from the intestinal lumen (Smith, 1967). Changes in intestinal membrane permeabilities can lower or stop the passive influx of water for such fish. In the case of sea water fish, the ingested water is transported by passive flow following an active uptake of ions at the level of the intestine. This water is retained while at other sites the ions are actively transported.

The skin of fishes is relatively impermeable to both water and ions. However, water loss in marine species and water influx across the integument of freshwater fish are known to occur. Transintegumentary ion fluxes also occur in aquatic and marine fishes. Mucous secretions and scales contribute to this low permeability but in the case of the former, a site for osmotic and ionic regulation may exist.

Hormonal Control of Osmo(iono)regulation

The osmo- and ionic regulation of freshwater, sea water and particularly diadromous fish species, is closely associated with various hormonal systems (Table 5.1). The pituitary gland of fishes synthesizes and secretes a number of hormones (prolactin, corticotropin, growth hormone, thyroid stimulating hormone, arginine vasotocin) known to be involved directly or indirectly with osmo(iono)-regulation during fish migration. Other tissues such as interrenal, caudal neurosecretory system (urophysis), thyroid and corpuscles of Stannius, also play a role.

Prolactin. Prolactin was first implicated in fish osmoregulation when Pickford and Phillips (1959) demonstrated that the injection of ovine prolactin was the only preparation that caused the survival of hypophysectomized killifish, *Fundulus heteroclitus*, in fresh water. Subsequently, a wide variety of teleosts were shown to require their pituitary gland in order to survive in fresh water. Many investigations using these fish have demonstrated that prolactin was the only hormone, from among the many that were tested, that promoted survival

Table 5.1: Effects of various hormones on fish osmo(iono)regulation in different salinity environments.

Hormone	Action	Species	Reference
Prolactin	Survival of hypophysectomized fish in fresh water	*Fundulus heteroclitus*	Pickford and Phillips, 1959
	Inhibition by ergot alkaloids	Different species	McKeown, 1972; Kramer *et al.*, 1973; Brewer and McKeown, 1978
	Prolactin cell activity increases in fresh water	Many	see Schreibman *et al.*, 1973; Clarke and Bern, 1980
	Role of ectopic pituitary transplants	Many	Ball *et al.*, 1965; Olivereau, 1971; Chidambaram *et al.*, 1972; Nagahama *et al.*, 1974
	Na^+ retaining activity	*Poecilia latipinna* *Sarotherodon mossambicus*	Ensor and Ball, 1968 Clarke, 1973
	Gel electrophoresis — effects of low osmolality	Many	see McKeown and Brewer, 1978
	Blood levels in low osmotic environments	Pacific salmon	McKeown, 1970; McKeown and van Overbeeke, 1972; Leatherland and McKeown, 1974; McKeown and Brewer, 1978
	Purified extracts	Pacific salmon	Donaldson *et al.*, 1968; Idler *et al.*, 1978; Kawauchi *et al.*, 1983
	Prevents loss of gill Na^+	Many	see Dharmamba and Maetz, 1972
	Active gill Na^+ uptake in fresh water	*Fundulus kansae*	Fleming and Ball, 1972
	Decreased Na^+-K^+-ATPase in sea water	*Fundulus heteroclitus*	Pickford *et al.*, 1970
	Increase urine volume	*Carassius auratus*	Stanley and Fleming, 1967; Lahlou and Giordan, 1970
	Drop in urine osmolarity	*Gasterosteus aculeatus*	Lam and Hoar, 1967
	Increased kidney Na^+-K^+-ATPase	*Fundulus heteroclitus*	Pickford *et al.*, 1970
	Increased glomerular filtration rate	Many	see Clarke and Bern, 1980

Table 5.1 (Cont.)

Hormone	Action	Species	Reference
Prolactin (cont.)	Increased intestinal absorption of water in sea water	*Anguilla japonica*	Utida *et al.*, 1972; Hirano *et al.*, 1975
	Decreased urinary bladder fluid absorption in sea water	Many	see Hirano, 1975
	Bladder epithelium structural changes in sea water	*Gillichthys mirabilis*	Nagahama *et al.*, 1973
	Increased bladder Na^+ absorption	*Anguilla japonica*	Hirano, 1975
	Increased bladder Na^+-K^+-ATPase	*Anguilla japonica*	Utida *et al.*, 1974
	Skin, gill and intestinal mucus secretion	Many	see Clarke and Bern, 1980
	Osmotic permeability of gill	*Anguilla japonica*	Ogawa, 1974, 1975, 1977
	Osmotic permeability of urinary bladder	*Gillichthys mirabilis*	see Doneen, 1976
	Osmotic permeability of intestine	*Anguilla japonica*	Utida *et al.*, 1972
	Osmotic permeability of kidney tubules	*Platichthys stellatus*	Foster, 1975
Corticotropin	Corticotroph hypertrophy during adult anadromous migrations	Pacific salmon	see Woodhead, 1975
Interrenal steroids	Reduced renal filtration, urine volume and extrarenal Na^+ uptake in fresh water	*Anguilla anguilla*	Chan *et al.*, 1967, 1969
	Stops gill Na^+ uptake in sea water	*Anguilla anguilla*	Mayer *et al.*, 1967
	Na^+-K^+-ATPase activity increase in gills, intestine and kidney	*Anguilla anguilla*	Pickford *et al.*, 1970; Epstein *et al.*, 1971
Vasotocin	Diuretic	*Anguilla anguilla* *Carassius auratus* *Protopterus aethiopicus*	Sawyer, 1972
	Antidiuretic	*Anguilla anguilla* *Oncorhynchus kisutch*	Henderson and Wales, 1974 McKeown *et al.*, 1976
Urotensins	Increased activity of caudal neurosecretory system upon salinity change	*Sarotherodon mossambicus* *Salvelinus fontinalis* *Mollienesia sphenops*	Fridberg and Bern, 1969 Audet and Chevalier, 1981 Kriebel, 1980

Table 5.1 (Cont.)

Hormone	Action	Species	Reference
Urotensins (cont.)	Increased plasma Na^{+}, Cl^{-} and osmolality	*Ophiocephalus maculatus*	Woo *et al.*, 1980
		Gillichthys mirabilis	Fryer *et al.*, 1978
	Water and NaCl transport across the intestine	*Ictalurus punctatus*	Richman and Barnawell, 1978
		Gillichthys mirabilis	Mainoya and Bern, 1982
	Cl^{-} transport in jaw skin	*Gillichthys mirabilis*	Marshall and Bern, 1979a, 1979b
	Na^{+} transport in urinary bladder	*Gillichthys mirabilis*	Loretz and Bern, 1981
	Diuresis	*Anguilla anguilla*	Chan, 1975; Chan and Bern, 1976
Corpuscles of Stannius	Hypocalcemia	*Anguilla japonica*	Chan, 1972
		Anguilla rostrata	Fenwick, 1974
	Gill Ca^{++} transport	*Anguilla rostrata*	Fenwick and So, 1974; Fenwick, 1976; So and Fenwick, 1977
	Glomerular filtration rate	*Anguilla anguilla*	Chester-Jones *et al.*, 1966; Sokabe *et al.*, 1970
Calcitonin	Hypocalcemia	Pacific salmon	Copp *et al.*, 1962
Thyroid hormones	Sea water preference	*Gasterosteus aculeatus*	Baggerman, 1957, 1959, 1960, 1962, 1963
		Pacific salmon	McInerney, 1961, 1964
Epinephrin	Gill water and ion transport	Many	Pic *et al.*, 1973
Renin-Angiotensin	Kidney function	*Salmo gairdneri*	Brown and Oliver, 1983

in fresh water (see Clarke and Bern, 1980). Ergot alkaloids have been used to block prolactin release in fish in place of surgical removal techniques. These studies have also shown that prolactin is required for a number of different species to survive in fresh water (McKeown, 1972; Kramer *et al.*, 1973; Brewer and McKeown, 1978) (Figure 5.3). Although there are about one dozen different

Figure 5.3: Cumulative mortality in rainbow trout in dilute sea water following daily injections of ergocryptine.

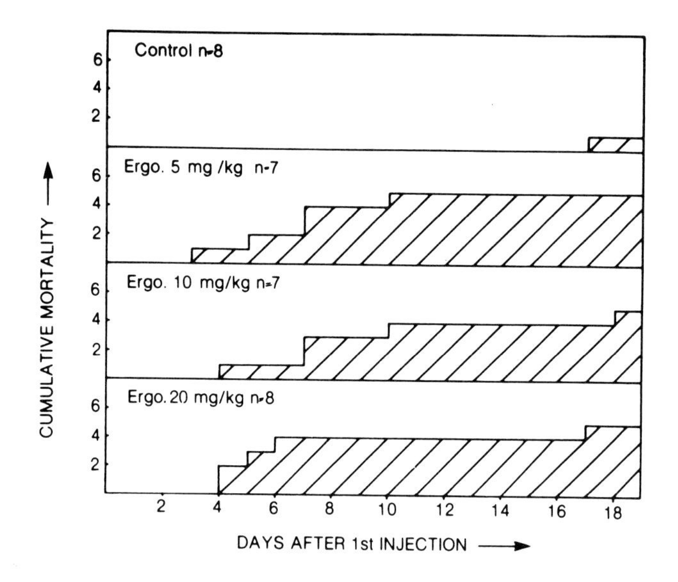

Source: from Brewer and McKeown, 1978

species that are described to be prolactin-dependent for survival in fresh water, there are an equal number of other species in which prolactin removal is not absolutely critical. Many of these latter species do however display ionic and osmotic imbalances (see Clarke and Bern, 1980).

A large number of cytological studies *in vivo* as well as *in vitro* have correlated prolactin cell activity with environmental osmotic pressure (see Schreibman *et al.*, 1973; Clarke and Bern, 1980). When the environmental (or incubation media) osmolality decreases, there are signs of prolactin cell hypertrophy, hyperplasia, increased synthesis (Figure 5.4) and release (Figure 5.5). Further evidence for the role of prolactin in freshwater survival and osmoregulation in a number of fish species has been obtained from ectopic pituitary transplant studies (Ball *et al.*, 1965; Olivereau, 1971; Chidambaram *et al.*, 1972; Nagahama *et al.*, 1974). There have also been a number of investigations that have quantified pituitary and plasma prolactin levels in relation to osmotic pressure. Sodium retaining bioactivity of prolactin has been shown to decline when salinity

Figure 5.4: Prolactin cell follicles from Pacific salmon. Top four photographs are from fresh water acclimated coho salmon after receiving injections of tritiated leucine (a major amino acid found in salmon prolactin). Autoradiograms on right indicate labelling (possible prolactin synthesis) of prolactin cells (especially near lumen) and luminal bodies. Sea water acclimated coho salmon had lower grain densities. Bottom photograph shows prominant bodies in the lumen of prolactin follicles from adult freshwater sockeye salmon.

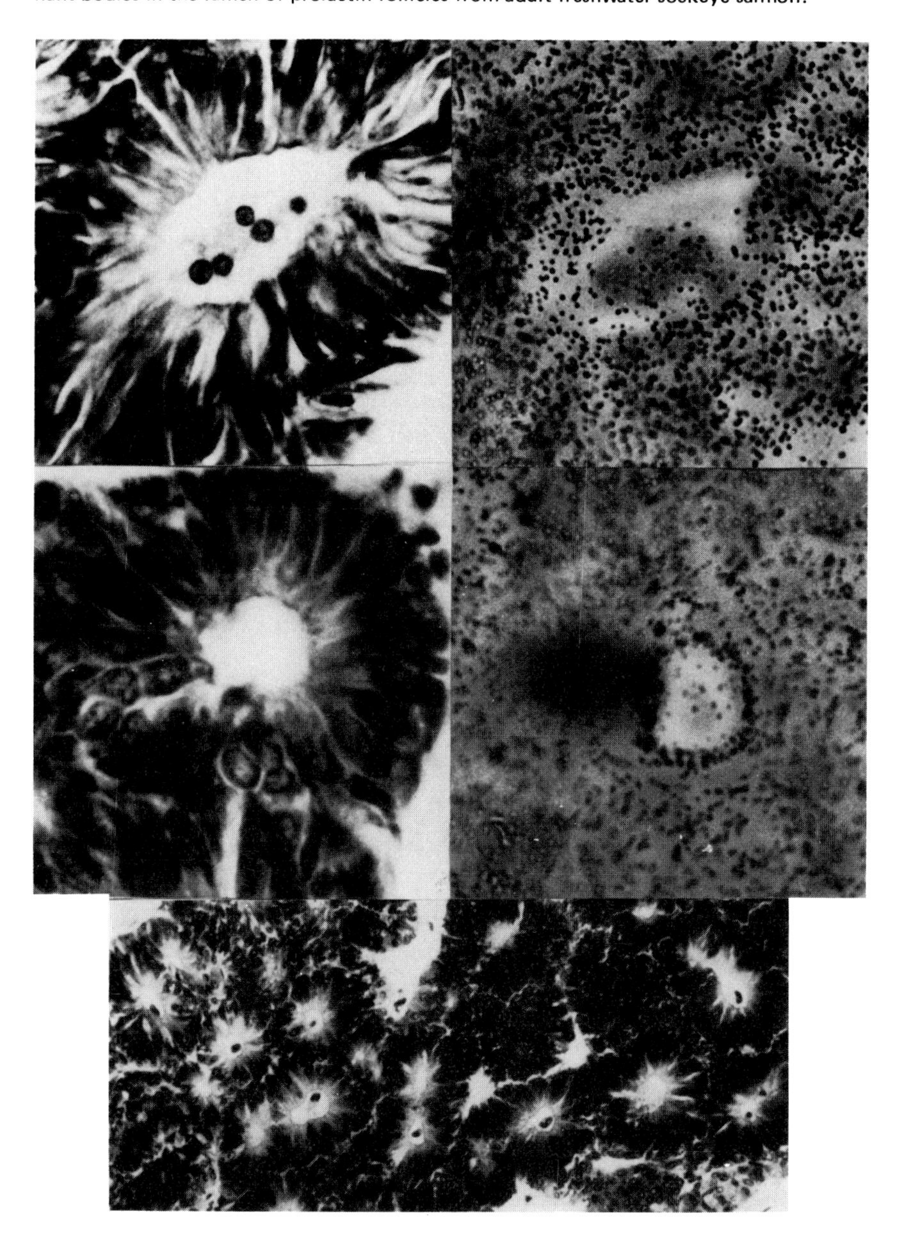

Source: from McKeown and Hazlett, 1975

Figure 5.5: Electronmicrographs of prolactin cells of fresh water acclimated kokanee salmon. (a) Whorl of endoplasmia reticulum; (b) base of prolactin cell adjacent to blood capillary; [c] note extensive endoplasmic reticulum, granule-release profile (arrow) and pinocytotic activity of the endothelial cells; (c) release of granules from prolactin cells base into the basement membrane complex; (d) peripheral part of a prolactin cell showing stages of granule release (arrows). Many small clear vesicles are near areas of exocytosis; (e) peripheral area of a prolactin cell showing granule release (arrow) into finger-like extensions of the basement membrane near pinocytotic vesicles [p] into the capillary endothelial cells.

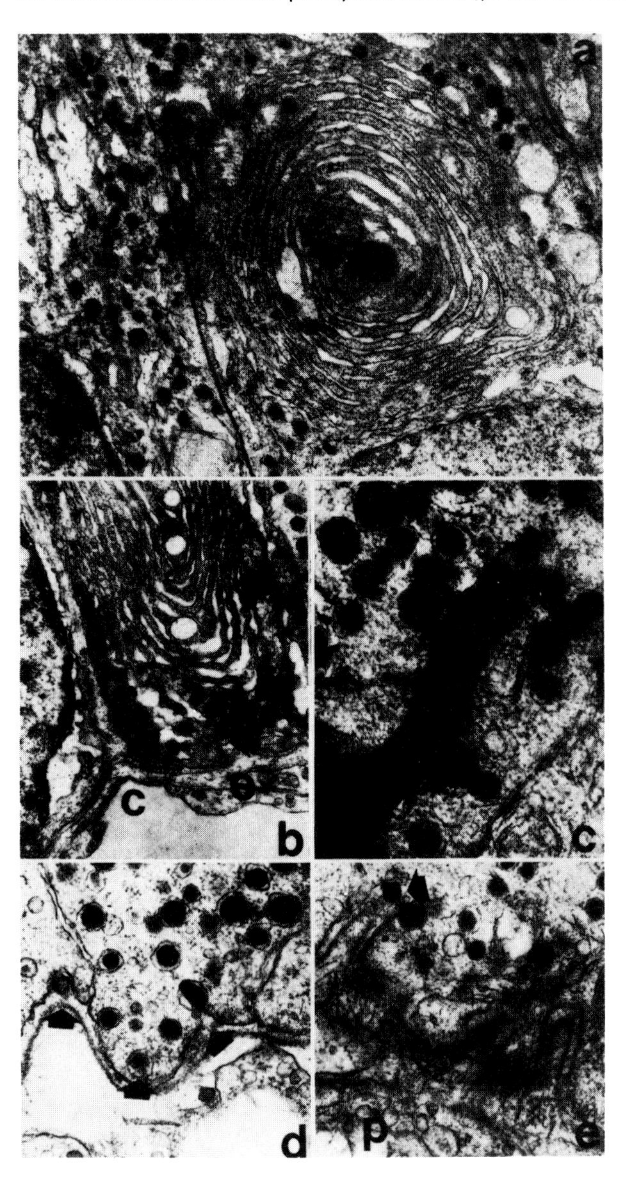

Source: from Leatherland and McKeown, 1974

increases (Ensor and Ball, 1968; Clarke, 1973). Quantification of gel electrophoretograms confirms that pituitary prolactin is reduced in sea water or when fish are transferred directly to fresh water (see McKeown and Brewer, 1978). This latter result suggests that fresh water is a stimulus for prolactin release and that the prolactin synthesis lags behind actual release. Concurrent measurements of both pituitary and serum prolactin in two different Pacific salmon species by heterologous radioimmunoassays further indicate that fresh water is a stimulus for rapid prolactin release (McKeown, 1970; McKeown and van Overbeeke, 1972; Leatherland and McKeown, 1974; McKeown and Brewer, 1978). Pacific salmon appear to possess a prolactin molecule that has bioactivity when tested for a fresh water osmoregulatory ability. An acid-acetone extract was thus shown to maintain the plasma osmolality of hypophysectomized goldfish in fresh water (Donaldson *et al.*, 1968). Isolated chum salmon prolactin also has sodium-retaining activity in hypophysectomized *Poecilia* (Idler *et al.*, 1978), and intact rainbow trout, *Salmo gairdneri*, adapted to 50 per cent sea water (Kawauchi *et al.*, 1983). Prolactin thus appears to be involved in adaptations to aquatic environments. However, pituitary prolactin levels are also high in fish in sea water or adapted to sea water (see Meier and Fivizzani, 1980). More work is required to ascertain whether prolactin also plays an osmoregulatory role for fish in the marine environment or whether it is involved in other processes such as metabolism or reproduction.

Prolactin influences both water and ion regulation and its site of action in fish includes gill, kidney, integument, urinary bladder and intestine. Many hypophysectomized fish in fresh water exhibit a passive loss of Na^+ via the gill epithelia which is decreased by prolactin (see Dharmamba and Maetz, 1972). Prolactin might also increase the gill active uptake of Na^+ in *Fundulus kansae* in fresh water (Fleming and Ball, 1972). Prolactin causes an elevated plasma Na^+ of fish in sea water due to a decreased gill Na^+ efflux (see Clarke, 1973) that might be due to a decreased gill Na^+-K^+-ATPase activity (Pickford *et al.*, 1970). There is evidence that prolactin affects ion regulation by the kidney since the drop in urinary volume caused by hypophysectomy is restored by prolactin treatment (Stanley and Fleming, 1967; Lahlou and Giordan, 1970). Lam and Hoar (1967) demonstrated that there was a drop in urine osmolarity of three-spine sticklebacks, *Gasterosteus aculeatus*, following prolactin treatment and at least in the killifish, this increased renal ion reabsorption was due to prolactin increasing kidney Na^+-K^+-ATPase activity (Pickford *et al.*, 1970). There is also considerable morphological evidence from light and electron microscopical studies of fish in fresh water indicating an effect of prolactin on kidney structure related to increased glomerular filtration rates and production of new nephric tubules (see Clarke and Bern, 1980). Euryhaline fish in sea water drink more and subsequently transport the ingested water and ions across the intestinal mucosa. Treatment with prolactin decreases this absorption (Utida *et al.*, 1972; Hirano *et al.*, 1975). Some marine but euryhaline species of fish exhibit an effect of prolactin on their urinary bladders. Prolactin treatment of these fish can cause a decrease in urinary bladder fluid absorption and a decreased water permeability of the bladder membrane (see Hirano, 1975). Morphological data indicate that the permeability changes are a result of structural alterations of the bladder epithelium as a result of prolactin treatment (Nagahama *et al.*, 1973). Prolactin is also known to increase Na^+ absorption across the bladder of these euryhaline marine fish (Hirano, 1975) and the hormonal effect is likely to be due to an increase in Na^+-K^+-ATPase activity (Utida *et al.*, 1974). A number of studies reviewed by Clarke and Bern (1980) have also implicated the integument as a site for prolactin action in freshwater fish. Mucous cells increase in numbers following prolactin treatment of both intact and hypophysectomized individuals. Other studies indicate increases in gill and intestinal mucous as well. Although this action of prolactin has been demonstrated in a number of species, many studies on other species have given negative results.

Prolactin has been shown to affect ionic regulation in a number of fish tissues in both fresh and sea water and in addition there are known effects on osmotic permeability. Fish in fresh water osmotically gain water and *in vitro* studies have shown that prolactin decreases the osmotic permeability of gill (see Ogawa, 1974, 1975, 1977), urinary bladder (see Doneen, 1976) and intestinal (Utida *et al.*, 1972) tissues. *In vivo* studies in the starry flounder, *Platichthys stellatus* (Foster, 1975), also suggest that there is a reduced kidney tubule osmotic permeability. It seems that the actions of prolactin in fish are so widespread that water and ion balance at most water-cell interfaces will be shown to be affected by this hormone.

Corticotropin and Interrenal Steroids. The effects of pituitary corticotropin and interrenal steroids on fish osmoregulation are considered here together since it is well established that interrenal activity in fish is dependent upon hypophyseal control (see Liversage *et al.*, 1971). Thus, the studies employing hypophysectomy or adrenalectomy techniques result in fish displaying similar osmotic and ionic imbalances. Likewise, replacement therapy with corticotropin or certain corticoid steroids leads to similar results (see Woodhead, 1975). The involvement of the pituitary-interrenal axis during fish migration with respect to osmoregulation is difficult to evaluate. Most research in this area has been done with diadromous species and at the time when the fish are changing from one salinity environment to another, when several different physiological processes are also being regulated. It is at this time that the migrating fish is mobilizing energy reserves for the physical activity of swimming and also using metabolic stores for gonadal maturation. Studies correlating corticotroph or interrenal cell structure and activity to osmoregulation functions are only circumstantial at best. Changes observed in interrenal cell activity from fish caught migrating solely in fresh or sea water suggest that the corticoid steroids are indeed involved with processes in addition to osmoregulation (Sterba, 1955; Robertson and Wexler, 1959; Honma and Tamura, 1963).

There are numerous investigations of anadromous fish species caught during migration indicating a stimulation of the pituitary-interrenal axis (see Woodhead, 1975). These studies include morphological observations of increased synthesis and release in both corticotroph and interrenal cells. Also, information has been obtained concerning the type of steroids released during spawning migrations, with cortisol and cortisone being the predominant hormones. Adrenalectomy of eels in fresh water caused a reduction of renal filtration rate, urine volume and extrarenal uptake of sodium, which was rectified by physiological doses of cortisol (Chan *et al.*, 1967, 1969). Mayer *et al.* (1967) also demonstrated that cortisol could rectify increased gill Na^+ uptake of adrenalectomized eels in sea water. It thus appears that cortisol is acting as a mineralocorticoid to regulate internal Na^+ levels in both fresh and sea water. Cortisol apparently transports Na^+ in the gills, intestine and kidney by stimulating Na^+-K^+-ATPase activity (Pickford *et al.*, 1970; Epstein *et al.*, 1971). It is interesting to note that aldosterone has similar effects to cortisol for freshwater eels although the large doses employed may be nonphysiological (Henderson and Chester Jones, 1967). There are, however, a number of conflicting results for the role of corticoids for osmoregulation in sea water (Woodhead, 1975). Ablation and replacement therapy experiments implicating a role of corticoids to promote salt loss in marine fish but salt absorption in freshwater species are substantiated by fresh to sea water transfer experiments. Thus, when fish are transferred to sea water plasma corticosteroids increase (Leloup-Hatey, 1964; Favre, 1960; Hirano, 1969; Ball *et al.*, 1971; Forrest *et al.*, 1973) and interrenal cells appear active histologically (Olivereau, 1962).

Vasotocin. Although vasotocin has not been determined to be present in fish blood, administration of this hormone results in osmoregulatory effects. Unlike in higher vertebrates, such fish as the eel, *Anguilla anguilla*, goldfish, *Carassius auratus*, and the lung fishes, *Lepidosiren paradoxa* and *Protopterus aethiopicus*, display a diuretic response to vasotocin injections (Sawyer, 1972). Diuresis is brought about by an increased glomerular filtration rate due to a constriction of efferent glomerular arterioles. However, when lower doses are employed the eel displays an opposite or antidiuretic response due to a decreased glomerular filtration rate (Henderson and Wales, 1974). Vasotocin may also be antidiuretic in the coho salmon, *Oncorhynchus kisutch*, at least just before it migrates to the ocean (McKeown *et al.*, 1976). Although studied in only a few fish species, vasotocin may be antidiuretic in the marine environment where fish need to conserve water but diuretic for freshwater species to eliminate excess osmotic water. In addition, vasotocin is known to increase active ion uptake across the gill in freshwater fish. It is possible that this action is mediated by the hormones' action on shunting blood towards areas rich in chloride cell populations (see Bentley, 1976).

Caudal Neurosecretory System. More than half a century ago large neurosecretory cells in the posterior portion of the spinal cord of fish were described

(Dahlgren, 1914; Speidel, 1919, 1922). The importance of this caudal neurosecretory system was not appreciated however until the work of Enami (1955). The neurosecretions from these cells are stored in a posterior neurohemal organ called the urophysis (Figure 5.6). The cells of the urophysis are strikingly

Figure 5.6: Caudal neurosecretory system in elasmobranchs and teleosts. The neurosecretions from the Dahlgren cells are transported to the neurohemal structure in the teleosts called the urophysis.

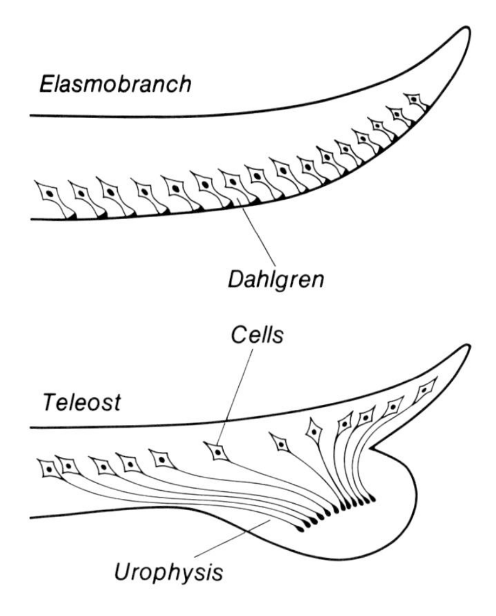

similar to those of the pars nervosa or posterior pituitary and were originally termed the urohypophysis. The blood supply from the urophysis enters the renal portal vessels via the caudal vein and then passes through the bladder and the kidney, making the urophysis ideally located for osmo(iono)regulation.

The protein neurosecretions stored in the urophysis which are termed urotensins have been isolated and characterized (Zelnick and Lederis, 1973; Lederis *et al.*, 1976; Lederis, 1977). These relatively small proteins, like the octapeptides of the neurophysis, are each associated with a larger distinct carrier protein termed urophysin (Masur *et al.*, 1976). There have been four different urotensins variously described as follows: Urotensin I, which decreases blood pressure when injected into rats; Urotensin II, which promotes smooth muscle contractions in such preparations as the trout urinary bladder and can increase eel blood pressure and urine flow; Urotensin III, which promotes the sodium uptake across the gills of goldfish (this effect has not been observed in other species); and Urotensin IV, which increases the *in vitro* water transfer across the toad urinary bladder. Urotensin IV may be identical to arginine vasotocin.

One of the first functions proposed for the caudal neurosecretory system was a role in controlling osmoregulation in fish. Only recently however have the urotensins been unequivocally demonstrated to have direct effects on osmoregulatory processes (Loretz and Bern, 1981; Marshall and Bern, 1981; Loretz *et al.*, 1982). Evidence for an osmoregulatory role for the caudal neurosecretory system has come from three lines of investigation: (i) electrophysiological and ultrastructural changes in the caudal neurosecretory system following changes in salinity, (ii) changes in plasma osmolality and ion concentrations following urophysectomy or administration of urophyseal extracts and (iii) direct effects on ion transporting tissues by pure urotensins.

Electrical stimulation of the caudal neurosecretory system cells caused depletion of *Tilapia* urophyses and the cells responded to plasma sodium but not to plasma osmotic changes (Fridberg and Bern, 1969). Some cells increased their firing in response to an increased plasma sodium while a different cell population responded to lowered plasma sodium. This response was specific to sodium as other cations were ineffective. Also, Audet and Chevalier (1981) described enhanced monoamine stainability of neurons innervating the caudal neurosecretory organ cells of brook trout, *Salvelinus fontinalis*, upon transfer from fresh to deionized water. Similarly, Chevalier (1976) showed, on morphological grounds, greatly increased cellular secretory activity after transfer from fresh to deionized water in the same species. Transfer of brook trout to sea water resulted in a marked reduction in the staining ability of the monoaminergic fibres innervating the caudal neurosecretory system (Audet and Chevalier, 1981) and the neurosecretory cells themselves became less active (Chevalier, 1976). Further supporting evidence was obtained by Kriebel (1980) upon transferring mollies, *Mollienesia sphenops*, from fresh to sea water. He found a paucity of granules in the axonal terminals of the neurosecretory cells near capillaries and these vessels were more dilated and packed with erythrocytes. One hypothesis that arose from these studies is that the caudal neurosecretory system is responding to a change in salinity and not salinity *per se* (Chevalier, 1976). Future studies on the kinetics of the urotensins in the circulatory system should answer this question.

If, however, salinity changes are the triggering factors, it is attractive to postulate that the urotensins are short-acting, initial responses to an osmoregulatory burden, prior to the establishment of long-term control by other hormones. Should this be so, then one would expect that the urotensins would exert their effects by activating a pre-existing mechanism rather than by new machinery. Evidence that this might be the case, and that urotensins might act through prostaglandin mediation was given by Woo *et al.* (1980). Treatment with urophyseal extracts of the freshwater fish, *Ophiocephalus maculatus*, after acclimation to fresh water or 40 per cent sea water, led to increases in plasma Na^+, Cl^- and osmolality in the freshwater fish. The 40 per cent sea water acclimated group showed a decrease in these parameters whereas both groups exhibited an increased plasma glucose level. Pretreatment of the two groups with a prostaglandin inhibitor abolished the effect on plasma ion levels but the hyperglycemia still resulted. Similar plasma ion effects of administered urophyseal homogenates to sea water adapted *Gillichthys mirabilis* were found by Fryer *et al.* (1978). Such injections to intact and urophysectomized fish decreased plasma Na^+, Mg^{++} and Cl^- levels. It is interesting that no differences were noted between urophysectomized and sham-operated fish, lending credence to the theory of initial adaptation to salinity change being the osmoregulatory role for urotensins.

Other effects of urophyseal extracts showed an inhibition of radiosodium uptake by the posterior hindgut of the catfish, *Ictalurus punctatus* (Richman and Barnawell, 1978). Mainoya and Bern (1982) have recently studied the effects of UI and UII from *Gillichthys mirabilis* on *in vitro* intestinal absorption of water and NaCl in *Tilapia* adapted to fresh water or sea water. UI caused decreases in water and NaCl absorption in freshwater fish only, whereas UII increased water and NaCl absorption in sea water fish only. The results thus suggest that intestinal transport may be affected by UI in freshwater fish and UII in sea water fish. The effect of a purified urotensin has been shown by experiments with the isolated jaw skin from the sea water fish *Gillichthys mirabilis*. This preparation exhibits a short circuit current equivalent to the net efflux of $^{36}Cl^-$ (Marshall and Bern, 1979a, 1979b). The skin contains numerous chloride cells which have recently been shown to be unequivocally responsible for chloride excretion (Foskett and Scheffey, 1982). When purified urotensin II from *Gillichthys* or *Catostomus* is added to the skin preparation a marked decrease in the short circuit current and transepithelial potential difference is observed. Only a slight increase is caused by urotensin I and no effect by arginine vasotocin. A later study explored the effects of urotensin I on the isolated skin (Marshall and Bern, 1981). While purified urotensin I had only a slight but consistent effect on untreated skins, it was found that pre-treatment of the skin with epinephrine caused a large drop in the short circuit current, an effect which could be reversed by urotensin I. These studies used large doses of urotensins, perhaps larger than the *in vivo* jaw skin might be expected to receive. However, as the gills, kidneys and bladder would doubtless be exposed to larger amounts of hormone than the jaw skin, the major effects influencing ions would probably be exerted on these organs.

Loretz and Bern (1981) explored this possibility. The isolated urinary bladder of *Gillichthys* has a region that actively transports Na^+ to the serosa via an electrogenic sodium pump. Administration of purified urotensin II in physiological concentrations resulted in an increase in the short circuit current and thus the net uptake of Na^+. It is of interest to note that urotensin II exerted its effect after the preparation had been incubated for about two to three hours. The authors attributed this to *in vivo* exposure to endogenous urotensin II.

As *Gillichthys* is normally a sea water fish, living in a dehydrating environment, the actions of urotensin II on the bladder would have the effect of reducing the osmotic concentration of bladder contents. As this would increase the osmotic gradient between the body fluids and bladder contents, the net effect would be the passive reabsorption of water from the bladder. The absorbed Na^+ would presumably be excreted by the gills, as they are the main organ of monovalent ion regulation in sea water fish. However, the inhibition of Cl^- efflux through the jaw skin by urotensin II is contradictory to the postulate of a marine osmoregulatory role for urotensin II. It remains to be seen what effect urotensin II has upon the gill. That urotensin II inhibits *in vitro* release of prolactin from *Tilapia* (Pearson *et al.*, 1980) is consistent with a sea water role.

The osmoregulatory functions of urotensins are not limited to ion transport for they have also been shown to have renal diuretic effects (Chan, 1975; Chan and Bern, 1976). The administration of purified urotensin I and II to eels, *Anguilla*, led to vasopression and diuresis. Urotensin II was much more potent and had different renal effects. Urotensin I, while causing diuresis in dose-dependent fashion, did not alter inulin clearance rates until high doses, where the urine/plasma ratio for inulin decreased below one, indicating that urine volume actually exceeded the filtered load (Chan, 1975). At high doses of urotensin I the amount of Ca^{++} and Mg^{++} excreted exceeded the filtered load. Chan (1975) suggested that the differing effects of urotensin I and II on renal function would indicate that urotensin I would be of use to a sea water fish and urotensin II for a freshwater fish, contrary to their ion regulatory effects on the bladder. In any case, even though the actual roles played by the two urotensins are unclear, it seems well established that the caudal neurosecretory system does have an osmoregulatory function, seemingly in the initial adaption to a new salinity.

Corpuscles of Stannius. The corpuscles of Stannius are small endocrine bodies embedded in the kidney of most teleost fishes. When this tissue is removed from sea water adapted European eels, *Anguilla anguilla*, plasma calcium and potassium levels increase whereas there is a decrease in plasma sodium (Fontaine, 1964; Chan, 1972). Bentley (1976) has suggested that the Na^+ and K^+ changes are an indirect effect of cortisol brought about by the hypercalcemia. Injections of corpuscles of Stannius extracts into these stanniectomized eels lowers plasma Ca^{++} levels back to normal. Stanniectomy of freshwater Japanese eels, *Anguilla japonica*, caused Ca^{++} levels to drop in the urine whereas HPO_4^{--} levels increased (Chan, 1972). It thus seems possible that the corpuscles of Stannius are involved with lowering calcium levels of fish in sea water where environmental calcium concentrations are high (Pang, 1973). Fenwick (1974) has conducted experiments with the North American eel, *Anguilla rostrata*, that add credence to this hypothesis since he also observed hypercalcemia following stanniectomy although this symptom was dependent upon the presence of exogenous calcium. The corpuscles of Stannius appear to have an effect at the site of the gill in North American eels since there is an increase in net $^{45}Ca^{++}$ influx in perfused isolated individual gills (Fenwick and So, 1974; So and Fenwick, 1977) and an increase in the amount of extractable branchial Ca^{++}-activated ATPase (Fenwick,

1976). The corpuscles of Stannius may also alter kidney function since extracts of this tissue cause a drop in blood pressure of eels (Chester-Jones *et al.*, 1966; Sokabe *et al.*, 1970), which in turn may lower the glomerular filtration rate.

The principle in the corpuscles of Stannius that causes hypocalcemia has been termed hypocalcin (Pang *et al.*, 1974). Two cell types have been described in the corpuscles of Stannius in the Atlantic eel and the three-spined stickleback (Wendelaar Bonga and Greven, 1975). One of these cell types probably produces hypocalcin as it responds to changes in environmental calcium levels (Greven *et al.*, 1978). The second cell type was not found in two marine fish species and in eels was more active in freshwater adapted than sea water adapted fish. This latter cell type may thus also be involved in osmotic and/or ionic balance in fish.

Calcium may also be regulated in fish by another hypocalcemic factor found in the ultimobranchial bodies and termed calcitonin (Copp, *et al.*, 1962). This role of calcitonin, however, is equivocal since some researchers find no response in calcium levels of certain fish species (see Norris, 1980). The pituitary gland of fishes may be the source of a hypercalcemic factor since hypophysectomy causes hypocalcemia in eels and killifish (Pang, 1973). The pituitary factor may be prolactin which increases calcium levels in fish in a freshwater environment. There is also evidence that the pituitary factor, at least in cod and eels, is parathyroid hormone-like and has been called hypercalcin (Parsons, 1979).

Other Endocrine Factors. There have been many studies correlating increased thyroid gland activity with diadromous fish migrations (see Woodhead, 1975). Such increases have been shown for fish migrating from sea water to fresh water as well as the reverse but such associations need not necessarily be indicators of an osmoregulatory role for thyroid hormones. At the time of smoltification in salmonids there is an increase in the thyroid gland activity and there is evidence that this increase is involved with an osmoregulatory role (see following section on smoltification). There is evidence, however, that sticklebacks and Pacific salmon can sense salinity gradients and subsequently alter their behavior to seek out environments of higher salinity (Baggerman, 1957, 1959, 1960, 1962, 1963; McInerney, 1961, 1964). Thyroid gland activity appears to be involved with this salt water preference since such behavior can be altered by mammalian thyroid stimulating hormone, thyroxine or anti-thyroid compounds (Baggerman, 1960, 1962, 1963). The involvement of the thyroid gland in altering sensory physiology and activity patterns such as salinity preference will be discussed later in this chapter.

Many fish migrations occur during periods of gonadal maturation and at times when the fish are changing from one salinity environment to another. There have not been any experiments however that indicate a role of the gonadotropins or gonadal steroids in fish osmoregulation. Baggerman (1963) in fact showed that even salinity preference was not affected in sticklebacks by gonadectomy although the timing of the preference was altered. There are also a good number of examples of different fish species that migrate to water of different salinity when they are sexually immature or when their gonads are at various stages of maturation.

The epiphysis or pineal gland may affect fish osmoregulation indirectly by altering corticoid levels (Delahunty *et al.*, 1977). Pinealectomy of goldfish during the spring causes a reduction in circulating cortisol levels. The reverse was apparent if the fish were pinealectomized in the summer and exposed to a long photoperiod. Interestingly, pinealectomy had no effect in the summer on cortisol levels if the goldfish were acclimated to a short photoperiod.

Epinephrine is known to cause ionic and osmotic changes across the fish gill epithelium (Pic *et al.*, 1973). Such effects might be mediated by epinephrine's effects on vasoconstriction and thus shunting of blood away from exchange surfaces. It appears, however, that the observed effects of epinephrine are direct actions on branchial cell permeability and chloride cell activity. Epinephrine injections to fish in sea water reduces the gill active efflux of sodium and chloride. Branchial osmotic permeability is also increased in both fresh- and sea water adapted fish. Further, the effects of epinephrine on ion fluxes can be prevented by α-adrenergic blocking drugs whereas those effects on water movement are affected by β-adrenergic inhibitors. Catecholamines can increase in the circulation in response to a number of different stresses, accompanied by changes in water and ion levels. Whether the stress of migration and particularly the osmotic stress of moving from one salinity environment to another causes a release of epinephrine and/or other catecholamines remains to be ascertained.

Injections of corpuscles of Stannius extracts cause an elevation of blood pressure in teleosts which suggests the presence of a renin-like substance (Chester-Jones *et al.*, 1966; Sokabe *et al.*, 1970). Stanniectomy of North American eels, however, is not accompanied by blood pressure changes, although when placed in sea water, plasma renin activity increases. It is thus likely that a renin-angiotensin system exists in fish that might stimulate the release of mineralocorticoids which in turn would be involved with osmo(iono)regulation. Fish kidney tissue does contain renin in relatively high concentrations indicating that such a system might be operative. Brown and Oliver (1984) have recently demonstrated a role of the renin-angiotensin system in the regulation of trout kidney function.

The regulation of water and ionic concentrations in fish appears to be under a fine control by a multitude of factors. Fish that migrate from one salinity environment to another must therefore possess refined mechanisms for these adjustments. Most experimental evidence for such controls by necessity comes from controlled laboratory investigations of one or a few isolated control mechanisms. Unfortunately, few studies have been conducted with fish from their natural environments, especially diadromous species, and thus little information is available concerning the actual controls or sequence of events that take place in fish during adaptation to new salinity environments.

Smoltification

Many salmonid fish species are anadromous and migrate to the sea at very precise times of the year. Environmental cues initiate the onset of such migrations which may occur at different times in the life cycle of the fish depending on the species. Within the family Salmonidae there is a great diversity or spectrum for the existence of anadromous behavior. Some species are obligatory forms whereas others are adaptively or optionally anadromous. Yet other forms are potamodromous and thus migrate entirely within freshwater environments. The diversity and evolution of such anadromous behavior is well reviewed by Hoar (1976).

The onset of smoltification is brought about by environmental cues or stimuli (see Wedemeyer *et al.*, 1980). The main environmental parameter that initiates a smolt transformation appears to be rate of change of the daily photoperiod (Clarke and Shelbourn, 1977; Barbour and Garside, 1983). Temperature is also important and may either advance the stimulus of a changing photoperiod or in some species such as the steelhead trout, *Salmo gairdneri*, may be a direct control for the smoltification process (Wagner, 1974a, 1974b). Spring freshets also appear to advance photoperiod and/or temperature stimuli.

The transformation of a parr to a smolt is associated with a number of behavioral, morphological and physiological changes (Table 5.2). The feeding habits of the parr are curtailed and metabolism shifts to a catabolic function in response to a changing internal environment (Wedemeyer *et al.*, 1980). The consequence is a loss of body lipid stores with a concomitant decrease in the weight to length ratio (condition factor) (Lovern, 1934; Evropeytseva, 1959; Hoar, 1951; Woo *et al.*, 1978). The loss of such low density organic compounds is compensated for by an increased size of the swim bladder (Saunders, 1965). Although the loss of body lipids at the time of smoltification is well documented, the adaptive significance of such changes for the fish is not well understood. Hoar (1976) suggests that the lipids may be involved with growth and a channelling into protein, or they may be used as a fuel for the increased activity of the smolt since the rate of metabolism increases at this time (Baraduc and Fontaine, 1955). He further speculates that due to the qualitative change in the lipids (Lovern, 1934) there may be an alteration in membrane lipids which change their permeability characteristics in readiness for a move to sea water. Finally, he suggests that because the fats become more fluid (Brocklesby and Denstedt, 1933) there is a pre-adaptation of the fish for lower environmental temperatures.

At the time of smoltification fish display a general restlessness, almost eliminate territoriality and become semipelagic, feeding in the upper portions of the water column. Subsequently the fish exhibit schooling behavior as they enter streams towards their seaward migration (Hoar, 1939, 1965; Folmar and Dickhoff, 1980). In experiments, fish at this time also show a preference for water of higher salinity (Houston, 1957, 1960; Baggerman, 1959, 1962, 1963; McInerney, 1964).

Table 5.2: Changes associated with the parr-to-smolt transformations.

Process	Reference
Reduced feeding and a shift to a catabolic metabolism	Wedemeyer *et al.*, 1980
Loss of lipid and a decrease in the weight to length ratio (condition factor)	Woo *et al.*, 1978
Increased swim bladder size	Saunders, 1965
Increased metabolic rate	Baraduc and Fontaine, 1955
Restlessness, elimination of territoriality, become semipelagic and development of schooling behavior	Hoar, 1939; 1965; Folmar and Dickhoff, 1980
Preference for water of higher salinity	Houston, 1957, 1960; Baggerman, 1959, 1962, 1963; McInerney, 1964
Loss of parr marks and development of a silver colour	see Folmar and Dickhoff, 1980
Changes in fin coloration, head shape, reduction in skin mucus, scale attachment and appearance of chloride cells in the gill	Hoar, 1951
Various endocrine changes (see Table 5.3)	Conte and Wagner, 1965; Eales, 1965; Hoar, 1965
Osmo(iono)regulatory mechanisms for survival in sea water	Conte and Wagner, 1965
Changes in blood and other tissue ion concentrations	Malikova, 1959; Houston, 1960
Decline in renal excretory rates	Holmes and Stanier, 1966
Changes in liver glycogen and vitamins as well as body and serum proteins	Fontaine and Hatey, 1950; Malikova, 1959; Vanstone *et al.*, 1964; Koch, 1968; Sweeting, and McKeown, unpubl.
Visual pigment changes for a deeper and marine environment	Muny and Swanson, 1965

Smoltification is characterized by many morphological changes the most obvious of which is a loss of parr marks used as cryptic coloration patterns or social signals of juvenile status during earlier freshwater residence. The smolt becomes a silver color due to purine deposition beneath the scales and deeper in the dermis. There is also a change in guanine and hypoxanthine ratios in these layers (Johnston and Eales, 1967). The silvering of the smolt may be required entirely as an adaptive camouflage coloration for a pelagic fish. However, Hoar (1976) has suggested that it might also be essential for water balance in a hypersaline environment. The excretion of purine nitrogen requires water and by depositing purines instead, salmon would thus conserve water when in a dessicating environment. Other structural modifications include changes in fin coloration, head shape, reduction in skin mucus, scale attachment and appearance of chloride cells in the gills (Hoar, 1951). Various transformations that take place at the time of smoltification are influenced by various hormones and thus morphological changes take place at this time in different endocrine tissues such as in pituitary and thyroid cells (Conte and Wagner, 1965; Eales, 1965; Hoar, 1965).

The behavioral and morphological changes during smolt transformation are also accompanied by physiological and metabolic changes. Water and ion regulatory mechanisms are affected such that smolts have an increased resistance to sea water when challenged (Conte and Wagner, 1965). The ionic composition of the blood and other tissues also changes (Malikova, 1959; Houston, 1960). Such ionic changes as well as water content may be a result of a decline in renal excretory rates (Holmes and Stanier, 1966). Metabolic changes required for the smoltification process or ensuing migration occur with changes in liver glycogen and vitamin levels as well as changes in body and serum proteins (Fontaine and Hatey, 1950; Malikova, 1959; Vanstone *et al.,* 1964; Koch, 1968). Sensory physiology changes also take place as exemplified by changes in visual pigments (Muny and Swanson, 1965). Noting these documented changes, it will be interesting to follow research investigations in the future that attempt to elucidate the adaptive significance of such alterations.

Hormonal Control of Smoltification

Considerable research has been conducted implicating the activity of thyroid tissue as necessary for parr-to-smolt transformations. This work was initiated by Hoar (1939) who first observed changes in the thyroid of Atlantic salmon, *Salmo salar*, as they underwent smoltification. Shortly afterwards Landgrebe (1941) proposed that thyroid hormones were involved with this process. Subsequent histological examinations of fish from natural populations or examinations of changes following hormone administrations have added to this proposal (Piggins, 1962; Higgs *et al.*, 1977; Clarke and Nagahama, 1977; Bern, 1978). The acquisition of a silver coloration by smolts has been linked to thyroxine since ^{14}C-labelled glycine (a precursor to guanine) was localised in the skin following hormone treatment (Matty and Sheltawy, 1967; Sage, 1968). Whether guanine deposition is a direct action of thyroid hormones is difficult to assess at this time. Thyroid hormones in teleosts have been shown to influence lipid mobilization, carbohydrate metabolism and stimulation of protein synthesis (Gorbman, 1969; Folmar and Dickhoff, 1980). The by-products of such shifts in metabolism might thus indirectly appear as morphological changes such as guanine deposition.

Enhanced thyroid activity of smolts affects sea water osmoregulatory capabilities (Refstie, 1982). Gill microsomal Na^+-K^+-ATPase increases dramatically during smoltification when thyroid activity is also enhanced (Zaugg and McLain, 1976). Folmar and Dickhoff (1981) not only showed an increase in gill Na^+-K^+-ATPase of coho salmon during smoltification but also an increase in plasma T_4 and T_3 levels. Bern (1978) and Clarke and Nagahama (1977) have also obtained circumstantial evidence by thyroid histological studies following

Table 5.3: Changes in various hormones during the parr-to-smolt transformation.

Hormone	Action	Reference
Thyroid hormones	Increased thyroid activity	Hoar, 1939; Landgrebe, 1941; Higgs *et al.*, 1977; Clarke and Nagahama, 1977; Bern, 1978
	Silvering	Matty and Shetlawy, 1967; Sage, 1968
	Lipid mobilization, carbohydrate metabolism and protein synthesis	see Gorbman, 1969; Folmar and Dickhoff, 1980
	Sea water tolerance	Refstie, 1982
	Increased gill Na^+-K^+-ATPase	Zaugg and McLain, 1976; Folmar and Dickhoff, 1979, 1981; Powell and McKeown, unpubl.
	Change in visual pigments from porphyropsin to rhodopsin	Bridges and Delisle, 1974
Growth hormone	Sea water survival	Smith, 1956; Komourdjian *et al.*, 1976b; Clarke *et al.*, 1977
	Enhanced growth	Clarke and Nagahama, 1977; Bern, 1978
	Increased secretion during upstream adult migration	McKeown, 1970
	Glucogenesis and lipolysis	Leatherland *et al.*, 1974; McKeown *et al.*, 1975, 1976; Sweeting and McKeown, unpubl.
	Thyroid activation	Grau and Stetson, 1977; Milne and Leatherland, 1978
Prolactin	Decreased activity	Zambrano *et al.*, 1972; Bern, 1978
Interrenal steroids	Increased activity of interrenal cells	Olivereau, 1962, 1975; McLeay, 1975; Komourdjian *et al.*, 1976a
	Sea water osmo(iono)regulation: gill Na^+ excretion, water permeability and ATPase activity. Water and ion movements in the intestine and urinary bladder	See Folmar and Dickhoff, 1980
	Stress reaction	Hirano, 1969; Ball *et al.*, 1971; Forrest *et al.*, 1973
Corpuscles of Stannius (Hypocalcin)	Hypocalcemia	Bern, 1978
Islets of Langerhans	Increased endocrine pancreas cell activity	Clarke and Nagahama, 1977; Bern, 1978
Insulin	Increased insulin release and lowered blood glucose levels	Patent and Foa, 1971; Woo *et al.*, 1978
Caudal Neuro-System (Urotensin II)	Osmo(iono)regulation	See Bern, 1978
	Increased arterial blood pressure	Chan, 1975

transfers of smolts to sea water for an osmoregulatory role of thyroid hormones in adapting fish to hyperosmotic media. Other work has demonstrated that thyroxine treatment was capable of regulating Na^+-K^+-ATPases that are required for sea water survival (Folmar and Dickhoff, 1979).

Another possible effect of thyroxine upon seaward migrating smolts is on vision. During freshwater existence, parrs specifically use the visual pigment porphyropsin. As smoltification proceeds the visual pigment changes to rhodopsin (Bridges and Delisle, 1974) which differs in its wavelength sensitivity. Rhodopsin is maximally blue sensitive (associated with deep water existence) and porphyropsin, red-sensitive (red light being more predominant at lesser depths). Changes in photoperiod in the spring may stimulate the hypothalamus via the eyes or pineal gland which in turn may activate the pituitary-thyroid axis to bring about visual pigment as well as other changes during smoltification.

The first evidence concerning growth hormone and its role in smoltification was given by Smith (1956). He found that injections of mammalian growth hormone enabled the brown trout, *Salmo trutta*, to survive salt water tests. Later studies revealed that injections of porcine, bovine, and teleost growth hormones also enabled Atlantic salmon parr and underyearling sockeye, *Oncorhynchus nerka*, to survive in salt water (Komourdjian *et al.*, 1976b; Clarke *et al.*, 1977). Further evidence by Clarke and Nagahama (1977), and Bern (1978) has shown a role for growth hormone in sea water adaption and growth. These experiments dealt with histological examination of coho salmon parr and smolt pituitary glands in both stunted and normal fish after release into sea water pens. The morphological changes coincided with the ability of fish to adapt to osmoregulation in a hypertonic environment. Smolts had a higher incidence of granulation and hypertrophied somatotrophic cells than normal fish. However, in nongrowing runts, somatotrophs still contain growth hormone although the secretion rate is unknown. This leads to the tempting conclusion that growth hormone has a function in addition to the promotion of body growth, that is, sea water adaption. This is supported by earlier studies by McKeown (1970) in which growth hormone levels were measured in both the pituitary and plasma following a gradual (Figure 5.7) or abrupt (Figure 5.8) transfer of sockeye smolts to sea water. Both types of transfers resulted in decreased pituitary growth hormone levels indicating probable release of the hormone into the blood. However, concomitant increases were not observed in circulating growth hormone levels. Instead, plasma somatotropin levels dropped shortly after a direct or step-wise transfer to sea water. It is likely therefore that the secreted growth hormone was utilized by some target organ for osmoregulation although the possibility exists that the secreted growth hormone was lost from the circulation because of breakdown or excretion at sites such as the liver and kidney. Such changes in hormone levels during salinity transfer experiments do not, however, elucidate the role the hormone has in osmoregulation, but only its changes in circulating kinetics. Growth hormone may thus only be involved in

Figure 5.7: Effects of step-wise transfers from fresh water to full sea water on pituitary and serum growth hormone concentrations in sockeye salmon smolts. Open circles — controls; closed circles — transferred fish. Arrows indicate a transfer to 1/3 sea water, 2/3 sea water, full sea water and fresh water in order from left to right.

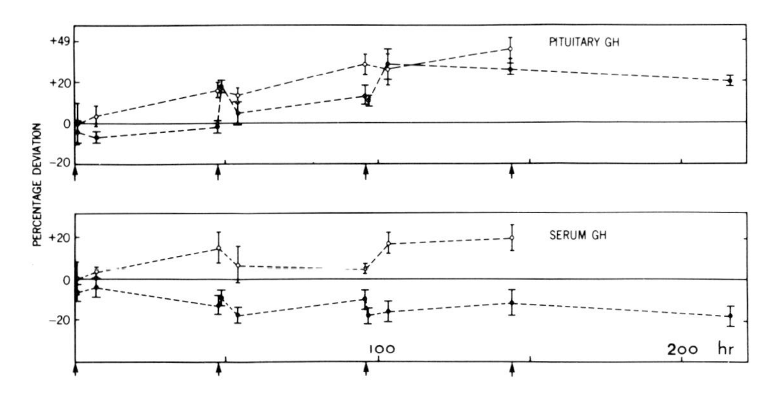

Source: from McKeown, 1970

Figure 5.8: Effects of an abrupt transfer of sockeye salmon smolts from fresh water to sea water on pituitary and serum growth hormone concentrations. Open circles — controls; closed circles — transferred fish.

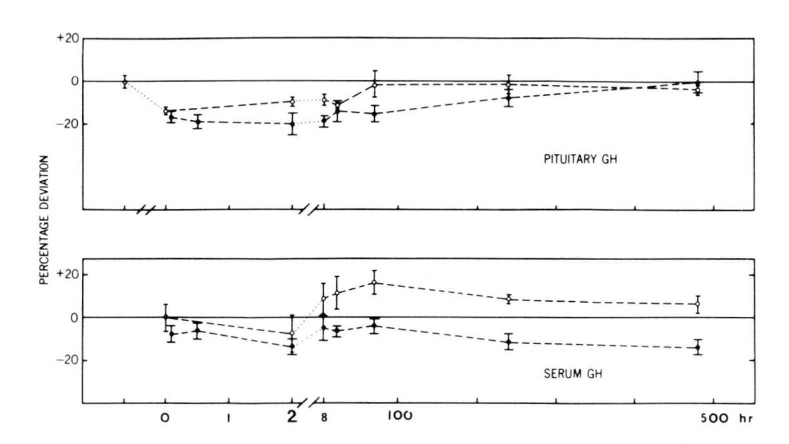

Source: from McKeown, 1970

osmoregulation indirectly, for example, by supplying organic compounds required for energetics for membrane transport mechanisms.

Although growth hormone may enhance sea water survivability of smolts, it may also be involved with changes in metabolic processes and general somatic growth. During normal salmonid growth there is an increased hypertrophy and hyperplasia of somatotrophs. Hypophysectomy of rainbow trout leads to a cessation of growth or at least a severely retarded growth (see Donaldson *et al.*, 1979). It also appears that growth hormone is glucogenic and lipolytic in certain Pacific salmon species (Leatherland *et al.*, 1974; McKeown *et al.*, 1975, 1976). Recent work by Sweeting and McKeown (unpublished) has shown a cyclical pattern to the growth of coho salmon during the parr-to-smolt transformation (Figure 5.9). Plasma glucose levels also fluctuate during this time. During specific times during this transformation, injections of either bovine or chum salmon growth hormone cause hyperglycemia (Figure 5.10). During smoltification of

Figure 5.9: Changes in wieght and plasma glucose of coho salmon during parr-to-smolt transformation.

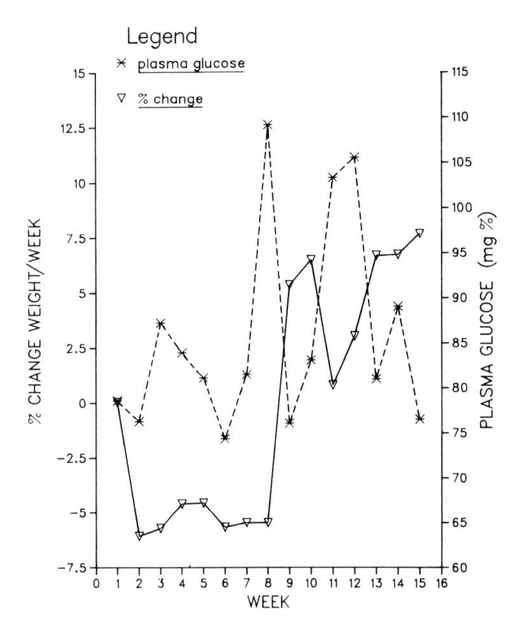

Source: Sweeting and McKeown, unpublished

coho salmon, growth hormone cells showed ultrastructural signs of enhanced activity when compared to parrs (Clarke and Nagahama, 1977). Coho that continued to grow in sea water exhibited enhanced somatatotroph cell activity whereas fish that failed to grow had lowered cell activity. This observation coupled with the fact that stunted fish had poor survival in sea water (Bern, 1978) leads to the suggestion that the change in condition factor brought about by growth hormone is required for survival of smolts in sea water. There is evidence that growth hormone affects the condition factor (weight to length ratio). Firstly, smolts weigh less per unit length than do parr (Wedemeyer *et al.*, 1980). Saunders and Henderson (1970) and Komourdjian *et al.* (1976a) found

Figure 5.10: Changes in plasma glucose concentrations in juvenile coho salmon following injections of saline (sal, control), bovine growth hormone (bGH), chum salmon growth hormone (sGH) and somatostatin (SRIF, somatotropin release inhibitory factor). Injections were administered at the beginning (Feb. 27), middle (Apr. 30) and end (Jun. 20) of the parr-to-smolt transformation.

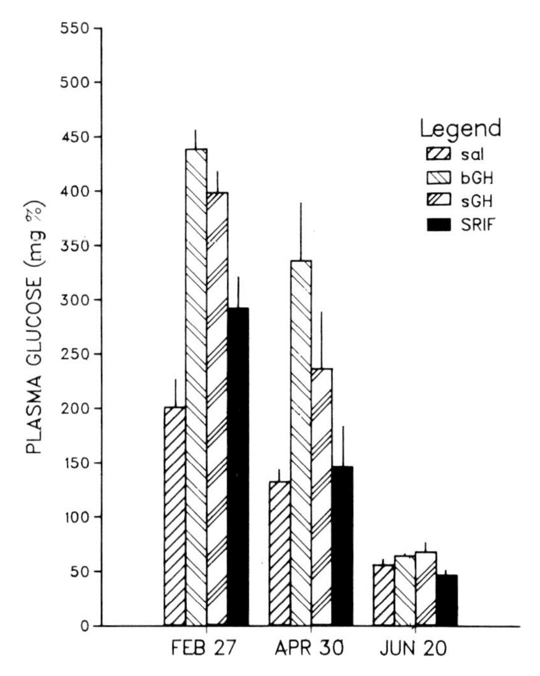

Source: Sweeting and McKeown, unpublished

that in natural stocks and after growth hormone injections that the condition factor of juvenile Atlantic salmon decreased.

At the time of rapid growth of parr, growth hormones appear to have an interplay with thyroid hormones. The term used to describe this interplay is the reciprocating thyroid-pituitary-light axis. The light aspect of this axis is achieved by the hypothalamic response to ambient photoperiodic influence (Hoar, 1965). Injections of ovine or bovine sommatotropin stimulate thyroxine (T_4) production in salmon (Grau and Stetson, 1977; Milne and Leatherland, 1978). Reciprocally, injections of T_4 show signs of hormone production in sommatotrophs (Sage 1967; Pandy and Leatherland, 1970; Leatherland and Hyder, 1975; Higgs *et al.*, 1976). Therefore, one of the effects of growth hormone, the secretory activity of which is probably enhanced by environmental stimuli such as photoperiod (Leatherland *et al.*, 1974), is to stimulate T_4 production and release during smoltification. This in turn is known to augment sea water survival. The secondary effect would then be to stimulate growth. Conversely, T_4 production in response to external cues may stimulate growth hormone activity to enhance growth, particularly as evidenced by the condition factor, which in turn increases marine adaptivity. Both reciprocal actions may of course be involved. By this axis, both sea water osmo(iono)regulatory capacity is achieved as is optimal size for smolt migration.

There is ultrastructural evidence that the prolactin-producing cells of coho salmon are less active in the smolt than in the presmolt (Bern, 1978). The role of prolactin in the smolting process, however, is as yet unknown; there seems to be an inhibition of prolactin secretion as smoltification and seaward migration progress. Experiments by Zambrano *et al.* (1972) on masu salmon, *Oncorhynchus masou*, indicated that neurosecretory innervation of the prolactin cells differed in the smolt when compared to the parr. This may suggest an increased hypophysiotropic activity in the smolt indicative of an inhibition of prolactin secretion preparatory to the fish entering sea water. Although prolactin cells appear less active in sea water smolts (Bern, 1978), the hormone is still present as was also shown for pituitary and plasma levels in salmon smolts transferred abruptly or step-wise into sea water (McKeown, 1970). It thus appears that prolactin physiology is altered at the time of smoltification but the hormone is probably still required for some function even after the smolt migrates into a hypersaline environment.

Activation of the pituitary-interrenal axis has been observed during smoltification of salmon (Olivereau, 1962, 1975; McLeay, 1975; Komourdjian *et al.*, 1976a). However, there is little information available concerning the physiological role of interrenal hormones in salmonid smoltification but there is evidence that cortisol in fish causes an adaption to a sea water environment (see previous section on osmoregulation). Cortisol thus increases Na^+ excretion, water permeability and ATPase activity of the gill as well as water and ion movements in the intestine and urinary bladder (see Folmar and Dickhoff, 1980). Cortisol levels increase in eels transferred to sea water, thus suggesting an osmoregulatory role in other fish, which may be coupled to the smoltification process in salmonids. This might only be a reaction to osmotic 'stress' since the cortisol appears to play a transitory role in sea water adaptation, as the elevations of plasma levels only last for a short time after transfer (Hirano, 1969; Ball *et al.*, 1971; Forrest *et al.*, 1973).

The corpuscles of Stannius secrete hypocalcin which, along with other endocrine factors, is involved with ion regulatory mechanisms (see previous section on osmoregulation). This tissue is probably involved with lowering blood calcium levels for fish in a hypercalcemic environment. Bern (1978) observed morphological alterations in cells of the corpuscles of Stannius indicating increased activity during smolt transformation of coho salmon. This may be a reflection of a preadaptation for calcium ion regulation following migration of smolts to the sea.

The endocrine pancreas or islets of Langerhans in salmonids is composed of a number of different hormone-producing cell types (Wagner and McKeown, 1981) (Figure 5.11). The hormones from this tissue are largely involved with homeostatic alterations in carbohydrate metabolism. At the time of smoltification

Figure 5.11: Immunocytochemical localization of different hormones in the endocrine pancreas of rainbow trout. (A) Clusters and cords of insulin cells, (B) peripherally located pancreatic polypeptide cells, (C) somatostatin cells surrounding insulin cells, (D) peripherally scattered glucagon cells surrounding insulin and somatostatin cells (arrow). Ex = exocrine pancreas.

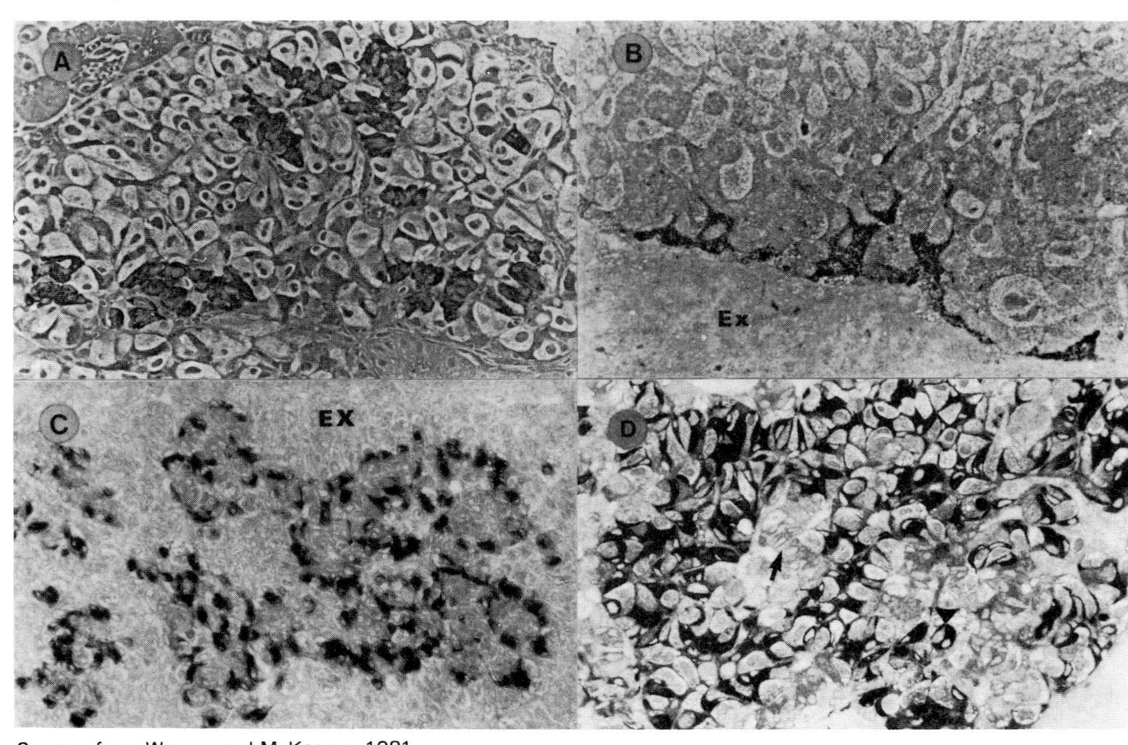

Source: from Wagner and McKeown, 1981

there are a number of energy demands with respect to morphological changes including general body growth and physiological alterations such as sensory mechanisms, behavior patterns and osmo(iono)regulatory preparedness for sea water. It is therefore not surprising that Bern (1978) observed increased hypertrophy of pancreatic islet cells of coho salmon smolts as compared to parrs. Clarke and Nagahama (1977) also observed signs of decreased cellular activity of islet cells in stunted fish at the time of parr-smolt transformations. At this time it is not known what cell type(s) in the endocrine pancreas is (are) stimulated in smolts. There are known differences, however, in circulating plasma insulin levels between normal and stunted fish (Patent and Foa, 1971). The insulin levels are lower in the stunted fish which might explain the observed higher blood-glucose concentrations of these fish (Woo *et al.*, 1978). The difference in insulin levels found between stunted fish and normal smolts might also be evident between smolts and parrs and thus the stimulation of pancreatic beta cells may be the cause of certain observed morphological changes such as body growth at the time of smoltification.

The caudal neurosecretory system of fishes has been discussed previously with respect to its role in osmoregulation. During the smoltification process there is an increase in the urotensin II content of the urophysis (Bern, 1978). This increased activity in the smolt may thus reflect a pre-adaptation to regulation in sea water during smolt migration. The increased urotensin levels might act directly on target cells to regulate water and ion concentrations during salinity transfer but might also act indirectly via other factors. Chan (1975) has observed that urotensin II can cause an increase in arterial blood pressure which in turn would supply the kidneys and bladder with more blood. In itself this might enhance transport mechanisms for both water and ions but it may possibly be involved also with regulating a response of the juxtaglomerular apparatus as an adaptation to the dessicating sea water environment. Obviously much more work is required to sort out these changes and their significance during the smoltification process.

Metabolism

Fish display a wide variation in migration patterns as well as in behavior and physiology during movements from place to place. Every species, however, requires energy sources and metabolic mechanisms in order to migrate successfully. The degree of such physiological adaptations of course varies. During migration energy may be required for a number of different purposes. At this stage in the life cycle of a fish, or at any other time for that matter, a certain amount of energy is consumed for general body maintenance. There are energy demands for an array of physiological mechanisms including muscle activity, nervous co-ordination, hormonal regulation, excretory processes, ingestion, assimilation, respiration, osmo(iono)regulation, etc. Imposed on top of these demands are energy supplies required for the long and often arduous journeys of migration. The exercise involved may be of uninterrupted duration until the migration is completed or may be sporadic depending on the demands and availability of energy resources. Nevertheless, the mobilization of energy from these resources is always required. A large number of fish migrations are directed towards spawning grounds. It is during these spawning migrations that gonadal maturation usually takes place. In the case of female fish, the energetic cost may be considerable. Diadromous species have a further demand on their energy resources when these fish change from one salinity environment to another. For example, the regulation of ion levels within a fish involves active transport mechanisms which are often opposite in direction. Energy is expended for these processes as well as for associated morphological changes, behavioral alterations and endocrinological controls.

The required energy supply for migration is obtained from the catabolism of organic compounds which are available directly through feeding activity or indirectly through stored body reserves. Many migratory fish meet their increased energy demands by increased feeding activity while other migrants, due to a variety of biotic and abiotic factors, rely heavily if not entirely on stored energy reserves. For the latter type of migratory fish, mechanisms become increasingly important for acquisition and storage as well as mobilization of energy-rich compounds at appropriate times. The preceding chapter dealt with the types of organic compounds that were preferentially accumulated in readiness for migratory activity as well as with sites within the fish used for premigratory deposition of energy reserves and the selective use of such compounds during migration. The immediate energy source in fish thus appears to be carbohydrates but most fish store only small amounts. Lipids and, to a lesser degree, proteins are the major storage compounds. However, during migratory activity, carbohydrate levels are maintained and in some cases even increased but do so at the expense of stored lipids and proteins. This chapter deals with the physiology of the deposition, and the mobilization of these compounds. The functional abilities of fish to store and subsequently use these fuels is largely controlled by the endocrine system. Changes in the metabolic rates brought about by any number of different mechanisms impinging on particular groups of organic compounds are also controlled or influenced by hormonal factors. It is only for organizational reasons that the physiological control of metabolism is separated from bioenergetics and similarly why the endocrine factors are isolated systematically (Table 5.4). Homeostatic regulatory mechanisms and the use of energy reserves are not separated by the fish but are intimately integrated one with another.

Growth Hormone

The involvement of growth hormone in fish metabolic processes has been well reviewed recently by Donaldson *et al.* (1979). Earlier studies have employed the use of mammalian growth hormone and a great deal of valuable information has

Table 5.4: Hormonal control of metabolism in fish during migration.

Hormone	Action	Species	Reference
Growth hormone	Growth	Many	see Donaldson *et al*., 1979
	Increased food conversion	*Oncorhynchus kisutch*	Markert *et al*., 1977
	Free fatty acid mobilization	*Oncorhynchus nerka*	Leatherland *et al*., 1974; McKeown *et al*., 1975; Clarke, 1976
		Oncorhynchus kisutch	Higgs *et al*., 1975, 1976, 1977; McKeown *et al*., 1976; Markert *et al*., 1977
	Protein anabolism	*Cottus scorpio*	Matty, 1962
		Salmo gairdneri	Matty and Cheema, 1976 cited in Donaldson *et al*., 1979; Enomoto, 1964
		Oncorhynchus kisutch	Higgs *et al*., 1975, 1976
	Diabetogenic	*Cottus scorpio*	Matty, 1962
		Salmo gairdneri	Enomoto, 1964
		Oncorhynchus kisutch	McKeown *et al*., 1976; Sweeting and McKeown, unpub.
		Oncorhynchus nerka	McKeown *et al*., 1975
	Increased insulin cell granulation	*Oncorhynchus kisutch*	Higgs *et al*., 1976, 1977
	Increased circulating levels during upstream migration and forced exercise	*Oncorhynchus nerka*	McKeown and van Overbeeke, 1972; McKeown *et al*., 1975
Prolactin	Lipogenic	*Fundulus chrysotus*	Lee and Meier, 1967
		Fundulus grandis	Joseph and Meier, 1971
		Oncorhynchus nerka	McKeown *et al*., 1975
		Notemigonus crysoleucas	Pardo and de Vlaming, 1976
	Diabetogenic	*Sarotherodon mossambicus*	Clarke, 1973

Table 5.4 (Cont.)

Hormone	Action	Species	Reference
Corticotropin	Increased granulation during upstream migration	*Oncorhynchus nerka*	McKeown and van Overbeeke, 1969
	Carbohydrate metabolism	*Ictalurus melas*	Chidambaram *et al*., 1973
Interrenal steroids	Increased synthesis and release during migration	Many	see Woodhead, 1975
	Increased activity during gonadal maturation	*Salmo gairdneri*	Robertson *et al*., 1961b
		Plecoglossus altivelis	Honma and Tamura, 1963
		Gadus morhua	Woodhead and Woodhead, 1965
	Gluconeogenesis	Many	see Norris, 1980
	Carbohydrate metabolism	*Anguilla anguilla*	Hatey, 1951; Chester-Jones *et al*., 1964
		Anguilla rostrata	Butler, 1968; Butler *et al*., 1969
		Ictalurus melas	Chidambaram, 1972; Chidambaram *et al*., 1973
Thyroid hormones	Protein catabolism	*Carassius auratus*	Hoar, 1958; Thornburn and Matty, 1963
	Protein anabolism	Many	see Donaldson *et al*., 1979
	Lipolysis	*Salmo gairdneri*	Barrington *et al*., 1961; LaRoche *et al*., 1966; Norris, 1969
		Salvelinus fontinalis	Naravansingh and Eales, 1975
	Carbohydrate metabolism	*Anguilla anguilla*	Fontaine *et al*., 1953
		Mugil auratus	LeRay *et al*., 1970
		Cyprinus carpio	Murat and Serfaty, 1971
		Salvelinus fontinalis	Hochachka, 1962
Androgens	Growth	Many	see Donaldson *et al*., 1979
Estrogens	Lipolysis	*Xiphophorus helerri*	Clavert and Zahnol, 1956
		Carassius auratus	Bailey, 1957
		Oncorhynchus kisutch	Ho and Vanstone, 1961; Vrist and Schjeide, 1961

Table 5.4 (Cont.)

Hormone	Action	Species	Reference
Insulin	Lipogensis	Many	see Meier and Fivizzani, 1980
	Protein anabolism	*Opsanus tau*	Tashima and Cahill, 1968
		Fundulus heteroclitus	Jackin and LaRoche, 1973
		Salmo gairdneri	Matty, 1975
		Esox lucius	Thorpe and Ince, 1974; Ince and Thorpe, 1976
		Anguilla anguilla	Ince and Thorpe, 1974
		Anguilla japonica	Inui *et al.*, 1975
	Hypoglycemia	*Opsanus tau*	Tashima and Cahill, 1968
		Esox lucius	Thorpe and Ince, 1974; Ince and Thorpe, 1976
		Salmo gairdneri	Wagner and McKeown, 1982
Glucagon	Hyperglycemia	*Channa punctatus*	Gill and Khanna, 1975
	Glycogenolysis	*Carassius auratus*	Birnbaum *et al.*, 1976
Vasotocin	Hyperglycemia	*Lampetra fluviatilis*	Bentley and Follet, 1965
	Hyperglycemia and lipolytic	*Oncorhynchus kisutch*	McKeown *et al.*, 1976
Catecholamines	Hyperglycemia and lipolysis	Many	see Meier and Fivizzani, 1980

been gained with respect to metabolic effects. More recently, growth hormone has been isolated, purified and characterized from a few fish species and comparative studies of its effects initiated. Growth hormones have an effect on general body growth and results are available indicating effects on specific target tissues and mobilization of certain organic compounds or conversion of various food materials. Some of these effects are apparent in fish at the time of smoltification, particularly with respect to weight-length relationships and adaptation to a sea water environment. The somatotropic hormone also affects food intake and appetite which will be discussed later in this chapter in connection with physiological effects related to behavioral alterations.

Pickford and her co-workers have done much of the pioneering work indicating that the pituitary contains a growth-promoting hormone. Hypophysectomy of the killifish is thus followed by body weight loss with variable affects on body length (Pickford, 1953, 1954, 1957, 1959; Pickford *et al.*, 1959; Swift and Picford, 1962, 1965). Most of these studies and many others by different investigators have demonstrated that the weight loss in hypophysectomized fish is due to the absence of growth hormone since replacement therapy or administration of purified mammalian somatotropins to intact fish cause growth (see Donaldson *et al.*, 1979). There is now experimental evidence indicating how growth hormone might influence the growth of fish. Markert *et al.* (1977) have demonstrated that the administration of bovine growth hormone causes an increased food conversion in yearling coho salmon. Growth hormone may thus cause food material to be more readily available for body growth which is of importance to fish as a means to premigratory conditioning. Equally important is the role of growth hormone in providing these same or similar materials for energy production during migration. Both these functions may be based on similar modes of action of growth hormone. In mammals growth hormone is known to affect lipid, protein and carbohydrate metabolism (see Turner and Bagnara, 1976). Similar biological activities are likely to occur in fish as well.

Injections of ovine growth hormone into fingerling kokanee salmon, *Oncorhynchus nerka*, caused an elevation of free fatty acids in muscle but an insignificant rise in circulating levels (McKeown *et al.*, 1975). Contrary to this finding, Higgs *et al.* (1975, 1976, 1977) and Markert *et al.* (1977) observed a decline in muscle lipid in coho salmon following bovine growth hormone treatment. In addition to different species, these studies employed different growth hormone preparations, sampled fish at vastly different times post-injection, and measured different lipid components. Also, none of these experiments investigated the effects of growth hormone on other body lipid deposits. Clearly more research is warranted to investigate the lipolytic or lipogenic action of growth hormone in fish.

The study by Clarke (1976) demonstrated a reduction in total body lipid of yearling sockeye salmon following growth hormone treatment. From this study it would appear that growth hormone is lipolytic. However, mobilization of lipid stores between body compartments and types of lipids involved are not known at this time. In support of a lipolytic action, plasma-free fatty acids in coho salmon fry appear to be elevated when plasma growth hormone increases (McKeown *et al.*, 1976) and in juvenile kokanee salmon circulating levels of growth hormone and free fatty acids vary according to a circadian rhythm (Leatherland *et al.*, 1974) indicating a possible relationship between the two blood borne factors. Unfortunately, the turnover rates and sites of mobilization are not known. The circulating lipids may be used to increase lipid stores but, depending on the energetic needs of the fish at the time, the lipids may instead be oxidized for energy liberation.

Protein anabolism seems to be stimulated in fish by administration of mammalian growth hormone. Matty (1962) observed a reduction in total plasma protein as well as plasma urea in the sculpin, *Cottus scorpio*, following hormone treatment. Subsequently, porcine growth hormone was shown to stimulate the incorporation of ^{14}C-labelled leucine into skeletal muscle protein of rainbow trout (Matty and Cheema, 1976 cited in Donaldson *et al.*, 1979). Total body protein also increases after mammalian growth hormone treatment of rainbow trout and coho salmon (Enomoto, 1964; Higgs *et al.*, 1975, 1976). As with lipids, the increased levels of proteins in the blood following growth hormone treatment maybe catabolised by the fish at times of high energy demands.

Growth hormone may be diabetogenic in fish since Matty (1962) found a hyperglycemic response in *Cottus scorpio* following hormone treatment. Enomoto (1964) also found glucosuria in rainbow trout which was in all probability preceded by hyperglycemia. In addition, circulating levels of growth hormone were found to be high in coho salmon when plasma glucose levels were also high (McKeown *et al.*, 1976). Injections of ovine growth hormone into kokanee salmon gave an insignificant rise in plasma glucose but did significantly increase liver glycogen, indicating a possible inhibition of peripheral glucose utilization (McKeown *et al.*, 1975).

There is now evidence available which suggests how growth hormone evokes its hyperglycemic response in fish. Higgs *et al.* (1976, 1977) found that bovine growth hormone administration to coho salmon caused an increase in size and numbers of the islets of Langerhans and an increase in the granulation within the insulin cells. They postulated that there was an increased insulin synthesis due to a number of possible actions of growth hormone. However, the observed increase in β-cell granulation might only have been an accumulation of insulin due to a decreased release. Thus, the effect of growth hormone to raise blood glucose levels might be brought about by an inhibition of insulin secretion.

Although there are no data available, the possibility exists that growth hormone may affect the endocrine pancreas in fish by causing the release of glucagon which in turn would lead to hyperglycemia. Growth hormone might initially act on target tissues besides the pancreatic islets and thus directly raise blood sugar levels. Such a direct action might be the inhibition of peripheral glucose utilization or increased food consumption. Such changes would thus affect blood glucose levels which in turn might be the stimulus influencing pancreatic hormone activity.

Somatotropin has also been purified from different fish species and displays growth promoting activity when tested in different fish. Growth hormone purifications from two piscine species – the hake, *Urophycis tenuis*, and pollack, *Pollachius virens* (Wilhelmi, 1955), and bioassayed in the killifish, exhibited growth promoting activity although to a lesser degree than bovine growth hormone (Pickford, 1954, 1957; Pickford *et al.*, 1959). Growth hormone prepared from the shark, *Prionace* sp., also promoted growth in the killifish and was more potent than the preparations from bony fish (Lewis *et al.*, 1972; Pickford, 1973). More recently, growth hormone has been purified from tilapia, *Sarotherodon mossambicus* (Farmer *et al.*, 1976), chum, *Oncorhynchus keta*, and coho salmon (Idler *et al.*, 1978; Wagner and McKeown, 1983a) and sturgeon, *Acipenser guildenstadti* (Farmer *et al.*, 1980), with bioactivity when tested in fish (Table 5.5). Interesting differences in bioactivity occur between

Table 5.5: Comparative amino acid composition of coho salmon, sturgeon, tilapia and ovine growth hormone.

	Coho[a]	Sturgeon[b]	Tilapia[c]	Ovine[d]
Lys	10.1	15.2	8.2	13
His	4.3	4.0	5.0	3
Arg	8.1	11.1	11.0	13
Asp	25.3	20.3	19.3	16
Thr	8.0	9.7	12.0	12
Ser	16.7	18.3	21.4	12
Glu	25.2	21.9	29.1	25
Pro	13.0	6.7	6.8	8
Gly	9.8	7.6	7.4	10
Ala	7.3	10.0	8.2	14
1/2 Cys	3.7	4.6	4.6	4
Val	9.3	9.3	6.0	7
Met	2.8	4.9	1.2	4
Ile	8.8	5.5	9.0	7
Leu	26.0	24.1	27.2	22
Tyr	5.4	6.3	7.2	6
Phe	7.1	10.5	6.7	13
Trp	0.6	1.0	1.0	1

Notes: a. Wagner *et al.*, 1984. b. Farmer *et al.*, 1980. c. Farmer *et al.*, 1976. d. Norris, 1980

the various fish growth hormones with the primitive fishes being more related to the mammalian growth hormones than the more advanced teleosts (see Wagner and McKeown, 1983). Even within the Pacific salmon, chum growth hormone appears to be equipotent to ovine growth hormone but less active than coho growth hormone (Wagner and McKeown, 1983; Wagner *et al.*, 1984). Future studies with these purified piscine somatotropins should provide more detailed information of the role of growth hormone for metabolic adjustments during fish migration.

Growth hormone is known to be released from the mammalian pituitary gland in response to a number of different internal and external stimuli (see Muller, 1974). One such stimulus is an increase in blood levels of energy-rich compounds such as certain amino acids, glucose or free fatty acids. Such conditions would occur during fasting. Another factor causing growth hormone release is stress, including physical exercise. These particular stimuli may be of interest with respect to fish migration since many fish stop feeding during migration and for those species that journey long distances, tremendous physical exercise is experienced. If growth hormone could be released under such conditions in fish, energy reserves could be mobilized at appropriate times for long arduous trips. Many spawning adult sockeye salmon undergo such long and difficult migrations and growth hormone may play such a role.

Pituitary and serum growth hormone has been measured from sockeye salmon at sea, upon entering the Fraser River, during upstream migration and on the spawning grounds (McKeown and van Overbeeke, 1972) (Figure 5.12).

Figure 5.12: Changes in pituitary and serum growth hormone concentrations in adult sockeye salmon migrating from the ocean to the spawning grounds at Chilko Lake, Canada. Numbers below group members indicate distance (miles) along migration route between collection sites and the mouth of the Fraser River (arrow) (-, sites at sea; +, in river). Shaded histograms represent percentage differences in lengths of fish (upper histograms) and serum protein concentrations (lower histograms).

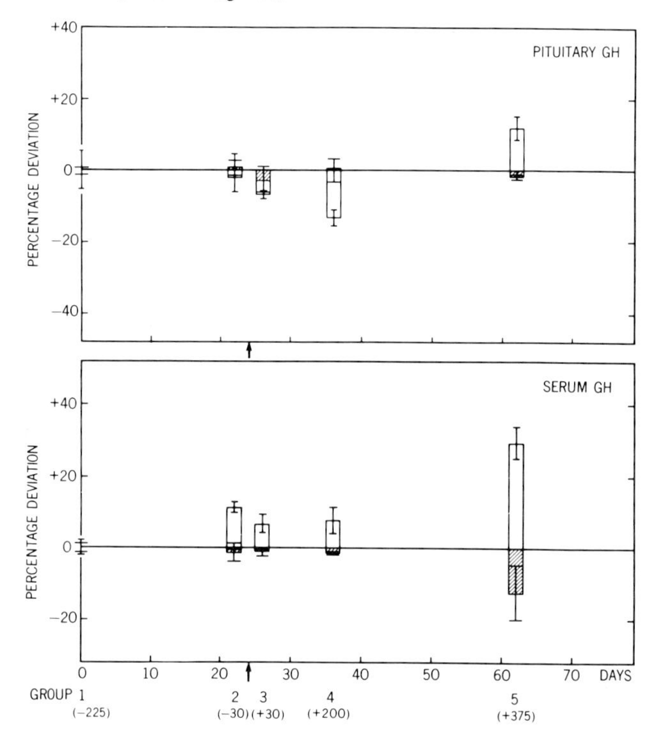

Source: from McKeown and van Overbeeke, 1972

Pituitary levels dropped during the upstream migration and were highest on the spawning grounds. This drop would be the time that the fish were actively swimming against the current and thus using their energy reserves. The drop in pituitary levels was probably a reflection of growth hormone secretion. During this same time of upstream migration the serum levels of the hormone remained unchanged which would suggest that the hormone was utilized from the blood, possibly because of metabolic demands. Measurements of growth hormone levels in fish taken from various locations near the fresh and sea water boundary indicated no changes (Figure 5.13). It would thus appear that the change to a

Figure 5.13: Changes in pituitary and serum growth hormone concentrations in adult migratory sockeye salmon as they approach the mouth of the Fraser River, Canada. For designation of symbols and numbers see Figure 5.12.

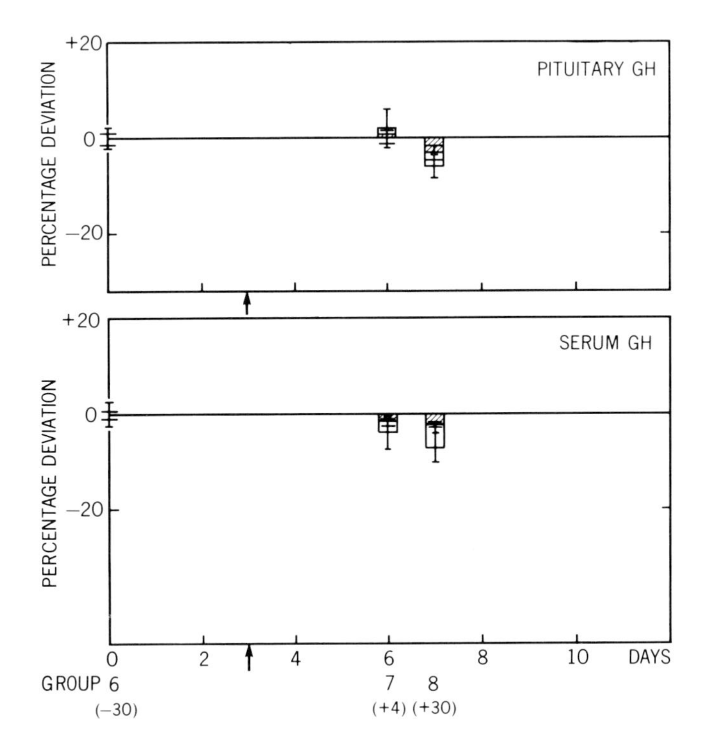

Source: from McKeown and van Overbeeke, 1972

lower salinity was not a stimulus for growth hormone release. However, it may be possible that growth hormone levels in the pituitary and serum were changing but were not observed since removal and replacement were equal. This seems unlikely since another study has also shown that growth hormone levels in kokanee salmon do not change under conditions of controlled environmental salinity alterations (Leatherland and McKeown, 1974). Since the study by McKeown and van Overbeeke gave only circumstantial evidence relating starvation and physical exercise to growth hormone secretion, controlled experiments were subsequently conducted (McKeown *et al.*, 1975). When kokanee salmon were deprived of food for 30 days plasma glucose tended to be lower and there was a significant reduction in plasma free fatty acids, liver glycogen and muscle water content. Plasma growth hormone levels were unchanged, however, but pituitary levels or turnover rates of the hormone in the circulation were not measured and the involvement of growth hormone during fasting in fish is thus still equivocal. In a second experiment kokanee salmon were forced to swim for 24 hours and compared to unexercised controls. The plasma growth hormone levels were significantly elevated in the exercised fish but various blood and tissue metabolite levels were unchanged. Growth hormone thus appears to be released during physical exercise but its subsequent role to mobilize energy reserves might not be manifested in only 24 hours.

Prolactin

A number of studies have indicated that prolactin is lipogenic in fish. Ovine prolactin has thus been shown to increase body lipid stores in *Fundulus chrysotus* (Lee and Meier, 1967), *F. grandis* (Joseph and Meier, 1971) and the kokanee salmon (McKeown *et al.*, 1975). Pardo and de Vlaming (1976) demonstrated that prolactin could directly stimulate hepatic lipogenesis *in vitro* in the golden shiner, *Notemigonus crysoleucas* although this could not be duplicated in another species (Meier and Fivizzani, 1980). Contrary to the above findings Clarke (1976) failed to observe a fattening response following prolactin injections into juvenile sockeye salmon. The negative results from this last study might be due to a daily variation in the fattening response to prolactin. The lipogenic response of a number of fish to prolactin has been shown to vary depending upon the time of day (Mehrle and Fleming, 1970; de Vlaming and Sage, 1972; de Vlaming *et al.*, 1975). The fattening response is restricted to a few hours after the start of a photophase and at other times of the day prolactin actually has an inhibitory effect. The phase of the response rhythm appears to be set by the onset of the daily photoperiod (Joseph and Meier, 1971). This response rhythm appears to be influenced by a temporal synergism of circulating corticosteroids (Meier *et al.*, 1971). Prolactin itself has also been shown to exhibit a circadian rhythm in circulating levels in a number of different fish species (Leatherland and McKeown, 1974; Leatherland *et al.*, 1974; Spieler and Meier, 1976a; McKeown and Peter, 1976; Spieler *et al.*, 1976, 1978). Daylength and temperature appear to modify these rhythms (Spieler, 1975; McKeown and Peter, 1976; Spieler *et al.*, 1978) (Figure 5.14). The control of the mobilization of energy reserves at specific times of the day by endogenous rhythms such as that for prolactin may thus govern daily activity patterns for migration.

There is little information available implicating prolactin as a metabolic hormone in fish other than its effects on lipid mobilization. Clarke (1973) is the

Figure 5.14: The effects of changing the photoperiod to 8 h light (8L): 16 h dark (16 D) for goldfish originally acclimated to a photoperiod of 16 h light: 8 h dark and at either 20°C or 10°C on pituitary and serum prolactin levels.

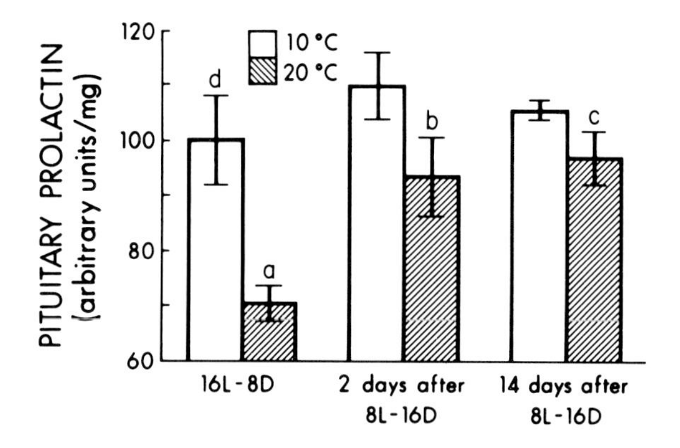

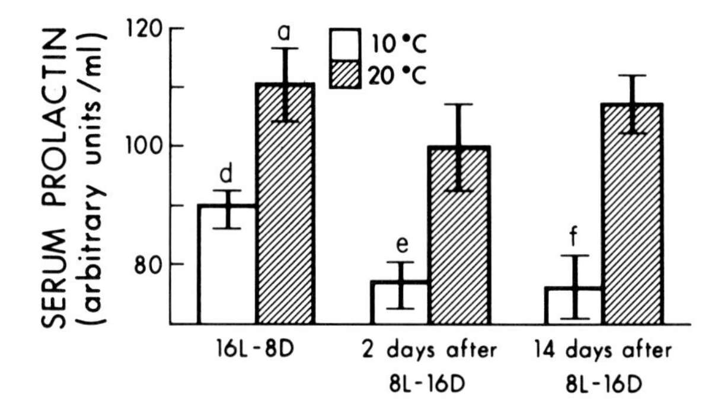

Source: from McKeown and Peter, 1976

only study I am aware of that demonstrates an effect of prolactin on carbohydrate metabolism. He showed that ovine prolactin caused a hyperglycemic response in *Sarotherodon mossambicus*. This result agrees with other studies in higher vertebrates where prolactin was found to also increase blood glucose levels as well as decrease liver glycogen stores (see Clarke and Bern, 1980). However, in a series of experiments by McKeown *et al.* (1975) there was no indication of an involvement of prolactin in carbohydrate metabolism in the kokanee salmon. Depriving these fish of food led to a reduction in liver glycogen and a tendency to lower plasma glucose but there was no change in circulating prolactin levels. Forced exercise did cause an increase in plasma prolactin but not a change in any parameters assessing carbohydrates. Such parameters were also unaffected by ovine prolactin injection. Although there may be differences in species hormone concentrations, route and number of injections, further work is required to delineate the role of prolactin in carbohydrate metabolism as well as possible involvement in protein mobilization.

Corticotropin and Interrenal Steroids

There have been a large number of investigations implicating a stimulation of the pituitary-interrenal axis during fish migration. Most of these studies have been conducted with the use of anadromous fish species and in particular salmonids. Since such fish change their salinity environments, it is difficult to ascertain whether the interrenal steroids are involved with osmo(iono)regulation or if they are mediating metabolic alterations for upstream migration. Mobilization of energy reserves may be initiated by glucocorticoids to supply energy for exercise, as well as to provide metabolites for gonadal maturation.

Evidence of an enhanced activity of the pituitary-interrenal axis during fish migration comes from a number of sources. The corticotropin producing cells in the pituitary gland of adult migrating sockeye salmon show an increased granulation during upstream migration although the secretion during this time is not known (McKeown and van Overbeeke, 1969) (Figure 5.15). Woodhead (1975) has produced an outstanding and comprehensive review of the changes that take place in interrenal activity during fish migration. There is morphological evidence of increased synthesis and secretion of adrenocorticosteroids by the interrenal cells of a number of different teleost species (Honma, 1959;

Figure 5.15: Microspectrophotometric analysis of granule density in ACTH cells of adult migratory sockeye salmon. Broken lines indicate the standard errors for each group. The group on the left were fish caught 45 km before entering the Fraser River, Canada. The two middle groups are river fish collected 60 and 300 km upstream. The group on the right was fish collected on the spawning grounds, 600 km upstream.

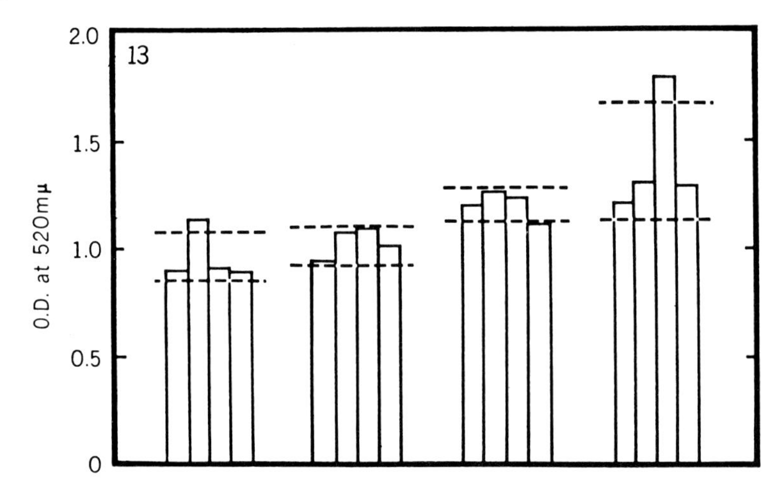

Source: from McKeown and van Overbeeke, 1969

Robertson and Wexler, 1959; Oguri, 1960; Donaldson and Fagerlund, 1970). Measurements of circulating levels of adrenocorticoids also add support to an activation of interrenal cell secretion. Studies of adult migratory Pacific salmon and Atlantic salmon indicate large increases of cortisone, cortisol and total 17-hydroxysteroids in the blood as the fish leave sea water and approach the spawning grounds (Hane and Robertson, 1959; Idler *et al.*, 1959; Robertson *et al.*, 1961a, 1961b; Leloup-Hatey, 1964). Migratory female salmon appear to have larger adrenocorticosteroid increases during the spawning migration than do males (Idler *et al.*, 1959; Robertson *et al.*, 1961a, 1961b; Hane *et al.*, 1965). This difference is likely to be due to a stimulatory effect of female gonadal steroids on interrenal cell activity (Fagerlund, 1967, 1970; van Overbeeke and McBride, 1971).

Robertson *et al.* (1961a) made interesting comparisons of plasma 17-hydroxycorticosteroid levels between spring-run and fall-run spring salmon, *Oncorhynchus tschawytscha*. The fall-run fish had much lower levels of circulating steroids and exhibited much shorter periods of starvation and their gonads were in a more advanced state of maturation. Thus, long periods of starvation requiring use of metabolic reserves for energy and gonadal maturation are correlated to high blood corticosteroid concentrations. Without knowing the turnover rates or kinetics of the circulating levels of adrenocorticosteroids, it is difficult to ascertain the reasons for the observed high levels during salmonid adult migrations. There are indications that such high levels might be due to hyperactivity of the interrenal tissue but other explanations include impaired steroid metabolism, reduced clearance or lowered plasma protein binding (Idler *et al.*, 1964; Hane *et al.*, 1965; Fagerlund, 1967; Idler and Freeman, 1967).

In order to differentiate between the roles of adrenocorticosteroids for metabolism and osmo(iono)regulation, studies of migratory fish solely in fresh or sea water become pertinent. Increases in activity of interrenal tissue have been observed in adult non-migratory rainbow trout (Robertson *et al.*, 1961b) and the freshwater salmonid, the koayu, *Plecoglossus altivelis* (Honma and Tamura, 1963) indicating a relationship between adrenocorticosteroids and gonadal maturation. A similar relationship was observed in the adult cod, *Gadus morhua*, in the marine environment (Woodhead and Woodhead, 1965). It is interesting to note that migratory adult rainbow trout exhibited a more pronounced interrenal tissue hyperplasia than non-migratory individuals (Robertson *et al.*, 1961b), again implicating an involvement of the pituitary-interrenal axis in the energy requirements of migration in addition to sexual maturation.

Cortisol and cortisone belong to a group of adrenocorticosteroids termed glucocorticoids and in mammals one of their functions is as a gluconeogenic agent. That is, they affect carbohydrate metabolism by supplying enzymes to convert proteins and lipids into carbohydrate precursors (see Norris, 1980). In addition, these steroids can alter membrane permeability to allow amino acids and lipids to be transported out of storage cells and taken up by carbohydrate synthesizing and storage tissues. Peripheral tissue glucose utilization also seems to be inhibited. These actions thus lead to hyperglycemia, increases in liver glycogen, increased urinary elimination and lipolysis. There is evidence that the glucocorticoids may act in a similar manner in fishes as well.

Although interrenal tissue in most fish is closely associated with kidney tissue, ablation surgery has been possible in American eels to ascertain the role of the piscine adrenal cortex homologue (Chester-Jones *et al.*, 1964; Butler *et al.*, 1969). Removal of this tissue in some eels leads to reductions in liver and muscle glycogen as well as lowered blood glucose levels. Hypophysectomy in many fish is followed by a depletion of liver glycogen and hypoglycemia. These symptoms are alleviated by replacement therapy of either adrenocorticotropin or corticosteroids (Hatey, 1951; Farkas, 1967; Butler, 1968; Tashima and Cahill, 1968; Pickford *et al.*, 1970; Chidambaram, 1972; Chidambaram *et al*;. 1973). Interrenal steroids may also similarly affect carbohydrate metabolism in elasmobranchs and cyclostomes (see Woodhead, 1975).

The physiological role of adrenocorticosteroids in fish may be of considerable importance during migration. The gluconeogenic action of converting protein and lipid stores into carbohydrate might be essential for energy demands during long arduous migrations and/or gonadal maturation. Evidence for factors stimulating the hypothalamo-pituitary-interrenal axis are of particular interest in this respect. The general adaptation syndrome described by Selye (1950) may apply to fish migrations. This adaptation in mammals involves the stimulation of adrenocortical hormone production and adrenal gland hypertrophy and hyperplasia. Due to the actions of the adrenocorticosteroids, an animal may be able to cope with extreme environmental conditions or 'stresses'. However, prolonged exposure to any extreme factor would provide the animal with required carbohydrates for adaptive measures but at the expense of body reserves of protein and lipid and eventually these stores would be depleted and the animal's defenses would be exhausted, so leading to its death. One particular environmental stress during migration of interest here is the continual exercise required for long distance migrants. The actions of interrenal steroids to provide carbohydrates as an intermediate energy source would be useful at such times during a migration. The fact that these hormones do increase during upstream migration of many adult salmonids adds credence to the general adaptation syndrome hypothesis. There is more direct evidence indicating that exercise is a stimulus to increase interrenal steroid secretion in fish. A number of experiments with different fish species have been conducted whereby the animals have been forced to swim and subsequently observed to exhibit elevated levels of circulating adrenocorticosteroids (Fontaine and Hatey, 1954; Fagerlund, 1967; Wedemeyer, 1969).

Thyroid

Thyroid hormones in mammals exhibit a number of known effects on metabolism (see Norris, 1980). They can thus cause the hydrolysis of fats or a so-called lipolytic action which, however, may only be a reflection of an action of other hormones potentiated by the thyroid hormones. Proteins also appear to be affected since the nitrogen balance is changed following thyroid hormone treatment. The effects on proteins can be either anabolic or catabolic depending on various factors such as different environmental parameters, stage of the life cycle of the animal and the particular tissue being considered. Thyroid hormones enhance carbohydrate metabolism and can lead to hyperglycemia. Glucose is oxidized faster under the effect of thyroid hormones and this is reflected in an increased basal metabolic rate as measured by an increased rate of oxygen consumption. This effect may be due to hormonal action to increase the synthesis of mitochondrial respiratory proteins such as cytochrome C, cytochrome oxidase and succinoxidase. If such actions of thyroid hormones were to be found in fish there would be an obvious metabolic advantage given to an actively migrating individual.

Thyroid hormones exhibit both protein anabolic and catabolic actions in fish. Thyroxine treatment in goldfish leads to an increased nitrogen excretion indicative of protein breakdown (Hoar, 1958; Thornburn and Matty, 1963). Smith and Thorpe (1977) observed an opposite effect in fed rainbow trout. This difference may be due to variations in the experiments such as temperature, for example, since Thornburn and Matty (1963) found differences in thyroxine-induced nitrogen excretion which were related to temperature. Although thyroid hormones appear to catabolize proteins in fish, *in vitro* as well as *in vivo* studies indicate an incorporation of specific amino acids into a variety of different body proteins among a variety of fish species following thyroid hormone treatment (see Dondaldson *et al.*, 1979). It may be that with increased absorption of digested proteins, the thyroid hormones are essential for the formation of new body proteins but under conditions of starvation, stored proteins may be broken down.

Evidence for a role of thyroid hormones to mobilize lipid reserves comes from radio-thyroidectomy experiments with rainbow trout whereby visceral lipid deposition was increased (LaRoche *et al.*, 1966; Norris, 1969). On the other hand trout treated with thyroxine displayed a decrease in abdominal fat (Barrington *et al.*, 1961) or reduced hepatic and visceral lipid reserves (Naravansingh and Eales, 1975). Such a lipolytic action of thyroid hormones in fish is not always supported experimentally (see Naravansingh and Eales, 1975).

Carbohydrate metabolism in fish is also affected by thyroid hormones. Thyroxine injections into European eels caused a reduction in liver glycogen stores (Fontaine *et al.*, 1953). A similar finding was observed by LeRay *et al.* (1970) for the mullet, *Mugil auratus*, after ingesting thyroxine but surprisingly, not after feeding with triiodothyronine (T^3). In the carp, thyroxine injections appear to increase carbohydrate storage in other tissues such as in cardiac and skeletal muscle (Murat and Serfaty, 1971).

If migratory fish can mobilize carbohydrate reserves from storage sites to active sites under the influence of thyroid hormone they would have a further advantage if these energy-rich compounds could be oxidized more readily. This may indeed be the case since Hochachka (1962) observed that thyroxine and T^3 both increased the rate of gluconate oxidation by brook trout liver preparations indicating the involvement of the pentose phosphate pathway. Also, cytochrome oxidase activity was increased in the mullet following treatment by thyroid hormones (LeRay *et al.*, 1970).

Gonadal Steroids

Vertebrates produce three main groups of gonadal steroids. The male testis produces the androgenic or testoid steroids with testosterone being a common example. The female ovary produces two types of steroids namely, the estrogenic or folliculoid steroids such as estradiol and the progestational or lutioid steroids such as progesterone. These steroid hormones influence vertebrate reproductive physiology for the development and maintenance of reproductive ducts and accessory organs and the development of secondary sexual characteristics including morphological as well as behavioral modifications. These actions will be considered in later sections of this chapter. In addition to these reproductive actions of the gonadal steroids, there is also evidence that in vertebrates, including fish, all three groups of these hormones also have an anabolic action.

Androgens. It is difficult to ascribe a general role of the male gonadal steroids to anabolic processes in fish as there seems to be a large variation of effects reported in the literature. Fish respond differently to androgen treatment depending on the species, sex, dosage, type of steroid and environmental factors (see Donaldson *et al.*, 1979). In treatment of the guppy, *Lebistes reticulatus* (Eversole, 1941; Scott, 1944; Clemens *et al.*, 1966), and the platyfish, *Platypoecilus maculatus* (Cohen, 1945), with androgens, the females exhibited a decreased growth but the males were unaffected. However, one study on the guppy showed an increase in growth of the females following testosterone treatment (Svärdson, 1943). A negative growth response was also observed in the brown trout (Ashley, 1957), following testosterone treatment but this response may have been due to the high doses employed. The negative growth response in some species may be due to a drop in body lipids since Pickford (1953) found that methyl testosterone in hypophysectomized killifish caused a depletion of liver lipid stores. Similarly, castration of the three-spined stickleback caused a deposition of body fat (Bock, 1928). Also, treatment of immature male sockeye salmon with 11-ketotestosterone caused an increase in body fat content (Idler *et al.*, 1961).

Contrary to the above studies there are a number of investigations that demonstrate an anabolic response of androgens in a number of salmonids, the tilapia, *Sarotherodon mossambicus*, and the goldfish (see review Donaldson *et al.*, 1979). Synthetic steroids such as 17 α-methyl testosterone which are low in androgenic actions can reportedly maximize the anabolic response. However, these studies also indicate that naturally occurring androgens such as testosterone and 11-ketotestosterone are also growth promoting.

Estrogens and Progestogens. The responses of different fish species to these two groups of steroids are rather variable (see Donaldson *et al.*, 1979). In some species and at certain doses they appear to promote growth whereas in different species opposite effects might be observed. However, studies that have investigated the effects of these steroids on specific tissues as opposed to total body growth, indicate that there may be an anabolic response in some tissues that might be at the expense of mobilizing reserves from other body areas. For instance, the estrogens are probably important metabolic hormones for mobilizing lipid stores to ensure yolk deposition in a developing ovary. Blood lipids are thus high in mature female salmon but not in males or immature fish and injections of estrogens into males or immature females increased the levels of circulating lipids (Ho and Vanstone, 1961; Urist and Schjeide, 1961). The goldfish shows a similar response to estrogen treatment (Bailey, 1957). Maturing adult female Barents Sea cod show increased levels of plasma lipids compared to males (Plack and Woodhead, 1966). These increased levels of lipids may be induced by estrogens during vitellogenesis and the source may be hepatic fat stores since Clavert and Zahnol (1956) observed a drop in liver lipids in the male swordtail, *Xiphophorus helerri*, following estrogen treatment.

Other Endocrine Factors

In higher vertebrates insulin functions as a hypoglycemic agent by converting circulating levels of glucose into stored glycogen or lipids. Insulin also stimulates incorporation of amino acids into protein and ingested lipids into fat stores. In many fish species insulin also appears to be lipogenic as it causes a drop in circulating free fatty acid levels (see Meier and Fivizzani, 1980). Tashima and Cahill (1968) demonstrated that insulin increased the uptake of labelled glucose into muscle protein and glycogen in the toadfish, *Opsanus tau*. In the killifish (Jackim and LaRoche, 1973), and the rainbow trout (Matty, 1975) insulin also promotes protein anabolism since hormone treatment caused labelled leucine incorporation. Circulating levels of amino acids are also reduced following insulin treatment in the Japanese eel (Inui *et al.*, 1975), the European eel (Ince and Thorpe, 1974) and the pike, *Esox lucius* (Thorpe and Ince, 1974; Ince and Thorpe, 1976). Insulin also seems to be hypoglycemic since hormone injections into the fish in the latter two studies caused reductions in blood glucose levels. Also, in a recent study by Wagner and McKeown (1982), higher circulating levels of insulin in rainbow trout were associated with lower blood glucose levels and higher liver glycogen concentrations. Some work has also been conducted which indicates that fish pancreatic islets also possess a hyperglycemic factor, glucagon. Injections of mammalian glucagon into Indian catfish, *Channa punctatus*, cause an increase in blood glucose levels (Gill and Khanna, 1975). Glucagon also causes *in vitro* glycogenolysis in goldfish hepatocytes (Birnbaum *et al.*, 1976).

Vasotocin may also be involved with mobilization of energy reserves in fish. Coho salmon injected with vasotocin exhibited increased circulating levels of glucose (Figure 5.16) as well as free fatty acids (Figure 5.17) (McKeown *et al.*,

Figure 5.16: The effect of vasotocin on plasma glucose in the coho salmon. S-30, 30 min post-injection with saline; S-120, 120 min post-injection with saline; 15V-30, 30 min post-injection with 15 mV vasotocin; 150U-30, 30 min post-injection with 150 mU vasotocin; 15V-120, 120 min post-injection and 15 mU vasotocin; 150V-120, 120 min post-injection with 150 mU vasotocin.

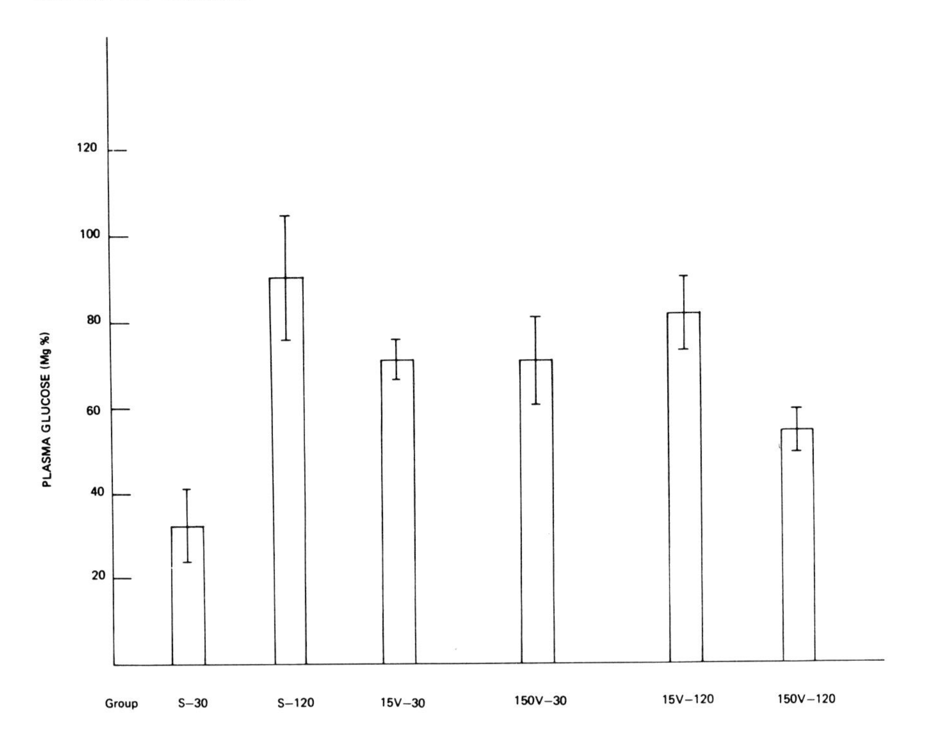

Source: from McKeown *et al.*, 1976

1976). Vasotocin also caused an elevation in plasma growth hormone (Figure 5.18) in these fish and thus the lipolytic and diabetogenic action of vasotocin might be mediated through growth hormone action. Bentley and Follet (1965) also found vasotocin to be hyperglycemic in the lamprey, *Lampetra fluviatilis*.

Epinephrine has been reported to be hyperglycemic in a number of different fish species (see Meier and Fivizzani, 1980). Catecholamines are also known to increase free fatty acid levels in the blood of certain fish (Larsson, 1973; Mazeaud, 1973). However, catecholamines do not cause lipolysis in adipose tissue when tested *in vitro* in the eel (Farkas, 1969) or rainbow trout (Fontaine, 1975). These hormones may thus act *in vivo* to stimulate free fatty acid release by indirectly stimulating some other factor(s).

Figure 5.17: The effect of vasotocin on plasma free fatty acids in the coho salmon. Symbols are the same as described in Figure 5.16.

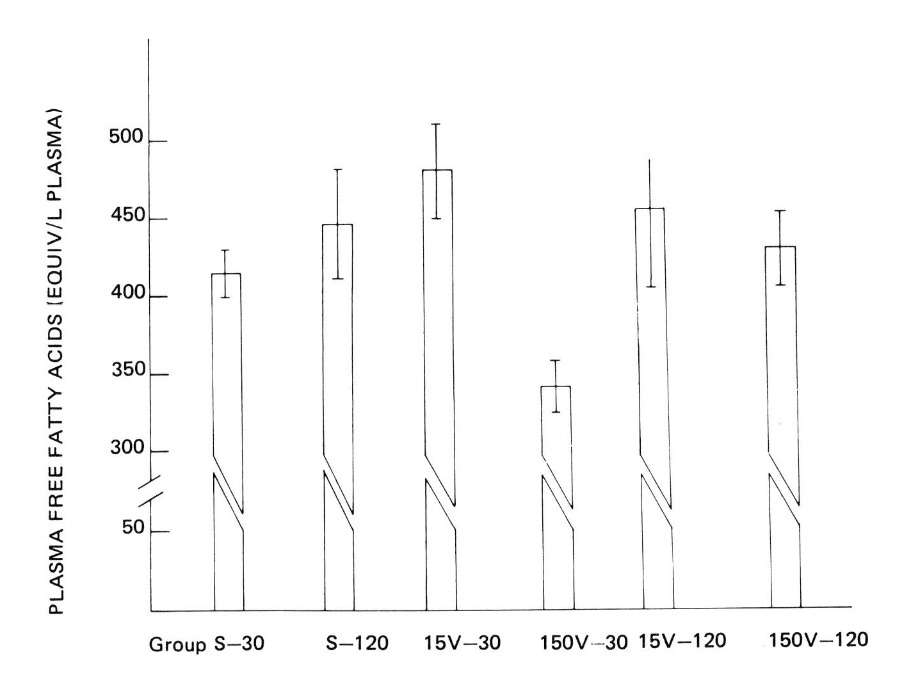

Source: from McKeown *et al.*, 1976

Gonadal Maturation

Aspects of reproduction in fishes have been extensively reviewed previously (Hoar, 1969; Liley, 1969; Yamamoto, 1969; Donaldson, 1973; Peter and Crim, 1979; Billard *et al.*, 1982; Crim, 1982; Demski and Hornby, 1982; Nagahama *et al.*, 1982; Peter, 1982). Fishes display a vast array of reproductive strategies ranging from such extremes as external to internal fertilization and from oviparity to true viviparity. During gonadal maturation different fish species undergo varying degrees of morphological, behavioral and physiological changes. Morphological changes of necessity take place in the gonadal tissue itself, but some species also assume drastic secondary sex characteristics and display a relatively high degree of sexual dimorphism for purposes such as mate selection, territoriality or aggressive behavior. At the time of gonadal maturation a number of physiological changes occur that control gonadal developmental changes in structure and function. Behavioral changes are also often controlled by internal secretions. Spawning migrations are sometimes preceded by gonadal development but more often sexual development takes place during migration or when fish reach the spawning grounds. The behavioral alterations that occur during

Figure 5.18: The effect of vasotocin on plasma growth hormone in the coho salmon. Symbols are the same as described in Figure 5.16.

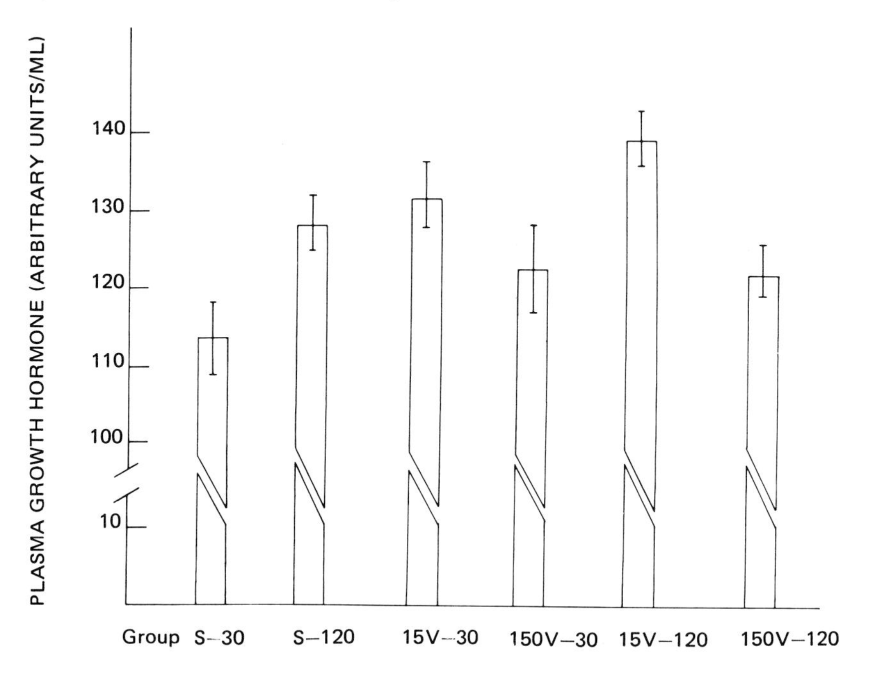

Source: from McKeown *et al.*, 1976

this time are such things as presexual behavior, courtship, mating and parental care. This section will not be concerned with the details of anatomical changes in gonadal tissue or other secondary sex characteristics as the topic is well reviewed in many other places. Similarly, reproductive behavior will not be a major concern here. The discussion here will however attempt to outline the controlling mechanisms for changes that take place with respect to reproduction related to a spawning migration. These physiological controls are mainly endocrinological and are initiated and modified by changing environmental parameters.

Reproductive Cycles

One of the major reasons why fish undertake extensive migrations is to find appropriate areas for spawning success whereby future generations are more likely to be sustained. Such migrations to spawning areas necessitate the concurrent development of gonadal tissues and appropriate behavior patterns. Thus, the development of migratory activity requires a timely environmental stimulus for the initiation of the journey and at the same time similar or related stimuli are required for reproductive development.

Some reproductive cycles appear to be endogenous, such as the circannual cycle in the female catfish, *Heteropneutes fossilis* (Sundararaj and Vasal, 1976). Even these rhythms, however, can be modified by various environmental factors. When a fish attains an appropriate internal environment, the external environmental stimuli of temperature and photoperiod seem to play key roles in initiating gonadal development (see Crim, 1982). Temperature regimes that stimulate reproductive events are often quite different for various fish species. There are, for example, a number of species such as the catfish (Sundararaj and Vasal, 1976), the sea perch, *Cymatogaster aggregata* (Wiebe, 1968) and the tench, *Tinca tinca* (Breton *et al.*, 1980a, 1980b) that display a gonadal recrudescence following an increase in temperature. Conversely, there are other species such as the marine fish *Mirogrex terrae-sanctae* (Yaron *et al.*, 1980) or the gobiid fish, *Gillichthys mirabilis* (de Vlaming, 1972) that require reduced temperatures to induce spawning in the winter. Gonadal recrudescence occurs in response to changing photoperiod but is as variable as the response of fish to temperature. Many fish species have gonadal maturation enhanced by lengthened photoperiods whereas others respond oppositely (see Peter and Crim, 1979; Crim, 1982). Many studies also indicate that there is an interaction between the two environmental factors and different species may be more dependent on one than the other. Also, during one phase of gonadal development, one environmental factor may play a major role whereas at another, a different cue may become important. Although temperature and photoperiod may be major features synchronizing gonadal recrudescence, the presence of other factors such as vegetation may also be important (Stacey *et al.*, 1979a).

The evidence is convincing that different environmental parameters are responsible for the timing of gonadal development and associated migrations. The environmental cues that bring about these changes in gonadal activity may do so because of an initial effect on the endocrine system. Crim (1982) has reviewed seasonal and annual cycles in changing hormonal levels that have been reported in fish. He summarizes changes that take place at all levels of the hypothalamo-pituitary-gonad axis. Thus there have been reported changes taking place seasonally in the hypothalamic content of gonadotropin releasing hormone in the carp, *Cyprinus carpio* (Weil *et al.*, 1975), the golden shiner, *Notemigonus crysoleucas* (de Vlaming and Vodicnik, 1975) and the tilapia, *Sarotherodon mossambicus* (King and Millar, 1980). Pituitary and plasma levels of gonadotropin vary seasonally in Atlantic salmon, carp, *Cyrpinus carpio,* rainbow trout and the tench, *Tinca tinca* (Billard *et al.*, 1978; Crim and Evans, 1978; Breton *et al.*, 1980a, 1980b). Surges in gonadotropin also appear to take place at specific times during the reproductive cycle such as the preovulatory surge of gonadotropin in the female goldfish (Stacey *et al.,* 1979a, 1979b). Daily or circadian rhythms in fish gonadatropin have also been described and are superimposed on the longer term seasonal rhythms (see Crim, 1982). Finally, gonadal steroids also respond to seasonal changes of gonadal development in the Atlantic salmon; rainbow trout; winter flounder, *Pseudopleuronectes americanus*; and the plaice,

Pleuronectes platessa (Idler *et al.*, 1971; Campbell *et al.*, 1976; Wingfield and Grimm, 1977; Dodd *et al.*, 1978; Whitehead *et al.*, 1978; Scott *et al.*, 1980).

In addition to gonadal activity cycling seasonally in response to circannual rhythms of sex hormones, the tissues of the hypothalamo-pituitary-gonad axis themselves may also vary in their responsiveness. The responsiveness of these tissues has been observed to change annually and indeed may do so in response to changing environmental parameters (see Crim, 1982). The reproductive cycles in fish are not only related to changing environmental parameters and fluctuations in hormones elaborated by the hypothalamo-pituitary-gonad axis, but also by other hormones such as cortisol and prolactin. Experimentation in the future might thus find that seasonal and/or daily changes in gonadal activity may be significantly affected by temporal synergism of many different hormones.

Reproductive Endocrinology

Little is known about how environmental cues stimulate the hypothalamo-pituitary-gonad axis in fish. However, there is some information available that the pineal gland and eyes are important since pinealectomy and blinding of goldfish suppresses gonadal activity and gonadotropin circadian rhythms (Vodicnik *et al.*, 1979; Hontela and Peter, 1980). Irrespective of how the environmental cues are responsible for stimulating gonadal activity much information is now available on the endocrine control of reproductive activity in fish. The hypothalamus has the highest level of control over gonadal activity and has been well reviewed recently by Peter (1982). Crude hypothalamic extracts from a number of fish species have been shown to contain a gonadotropin releasing hormone (GnRH) (see Peter and Crim, 1979). Chromatographic separation of hypothalamic peptides from the carp indicate that the molecular weight of the fraction with GnRH activity is less than 5000 (Breton *et al.*, 1975) which might indicate a comparison to luteinizing hormone-releasing hormone (LH-RH). King and Millar (1979, 1980) indicated that teleost GnRH is similar in size to that found in reptiles and birds but immunologically different from the releasing hormone found in amphibians or mammals. LH-RH and some of its synthetic analogues are known to stimulate gonadotropin release and induce ovulation in a number of teleosts (see Peter, 1982). The dosages of the releasing factor used in these studies often varies between studies and may be a reflection of different environmental conditions that augment the activity of the hormone. The type of environmental parameter that might alter the GnRH activity could be factors such as presence of vegetation (Stacey *et al.*, 1979a) photoperiod (de Vlaming and Vodicnik, 1975), pineal gland activity (de Vlaming and Vodicnik, 1977; Hontela and Peter, 1980) or circadian rhythmicity (Peter and Crim, 1978).

The location of the GnRH-producing cells has been investigated by a variety of techniques. Immunocytochemical investigations have demonstrated the localization of an LH-RH-like material in the pars nervosa of a number of teleosts (Dubois *et al.*, 1979; Nozaki and Kobayashi, 1979; Schreibman *et al.*, 1979). These observations are in agreement with other cytological studies with fish whereby neurosecretory axons have been observed making contact with gonadotrophs (see Peter, 1982). The immunocytochemical localization, however, of LH-RH nuclei is equivocal. Generally, the three different centers comprising the nucleus lateral tuberis, the telencephalon and the preoptic region, have been described in different fish species as being the centres for GnRH synthesis. Lesioning studies indicate the nucleus lateral tuberis as the source of GnRH in the goldfish (Peter and Paulencu, 1980; Peter *et al.*, 1980) and Atlantic salmon (Dodd *et al.*, 1978).

The release of pituitary gonadotropin in fish may also be controlled by a gonadotropin release-inhibitory factor (GRIF). Lesioning of the preoptic area causes an increase in serum gonadotropin in both male and female goldfish indicating the presence of a GRIF in that region (Peter *et al.*, 1978; Peter and Paulencu, 1980). Transplantation of the goldfish pituitary pars distalis to a position beside the brain of a recipient fish resulted in an increased serum gonadotropin level (Peter *et al.*, 1982). This presumably resulted from the transplant being deinhibited since a GRIF could no longer reach the transplanted pars distalis due to the disruption of normal *in situ* neuronal pathways. Transplantation of the pars distalis to the third ventricle of the preoptic region or to the ventricular space underlying the optic tectum resulted in lower levels of serum gonadotropin than transplants beside the brain. The transplants in the preoptic region are presumably in contact with some GRIF. There is evidence that the GRIF is dopamine. There is a decrease in the basal and GnRH-stimulated release of gonadotropin from the rainbow trout pituitary gland when treated with dopamine *in vitro* (Crim, 1982 cited in Peter, 1982). Dopamine or its agonist, apomorphine decreased the circulating levels of gonadotropin in unoperated goldfish (Chang and Peter, 1983). On the other hand, a dopamine antagonist had the opposite effect in unoperated goldfish. Dopamine or apomorphine also decreased serum gonadotropin levels in goldfish that had preoptic lesions which are known to normally increase gonadotropin levels due to a lack of a GRIF. Apomorphine also blocked the stimulatory action of an LH-RH analogue. Dopamine may thus be the GRIF which can act directly on the gonadotrophs and might do so by blocking the actions of GnRH.

Fish pituitary gonadotropin release is controlled by the hypophyseotropic factors and in turn the gonadotrophs control the physiology of the gonadal tissue. The existence of one or two gonadotrophs (see review Schreibman *et al.*, 1973) or gonadotropins (see reviews Donaldson, 1973; Peter and Crim, 1979; Burzawa-Gerard, 1982) in fish has been an ongoing argument for years. The reason for this controversy may have arisen from the fact that one of the fish gonadotropins is not a glycoprotein, unlike higher vertebrates (Campbell and Idler, 1976, 1977; Ng and Idler, 1978a, 1978b; Idler and Ng, 1979). Single glycoprotein preparations have been shown to elicit complete gametogenesis in hypophysectomized fish which would indicate only one gonadotropin being required (Billard and Escaffre, 1973; Donaldson, 1973). However, further purifications of salmon and tilapia gonadotropins have led to preparations that exhibit different specificities (see Peter and Crim, 1979). Also, Campbell and Idler (1976) and Campbell (1978) have described a nonglycoprotein pituitary fraction capable of vitellogenic activity.

The gonadotropic hormone is known to have a trophic action on the fish ovary directly as well as indirectly via certain steroid hormones. Gonadotropin appears to be responsible for the endogenous yolk formation in the ovary (Upadhyay, 1977). Oocytes grow by the uptake of vitellogenin which is synthesized by the liver. The synthesis and secretion of vitellogenin is influenced by estrogens which are in turn regulated by gonadotropin (Crim and Idler, 1978). The nonglycoprotein pituitary gonadotropin affects the oocyte uptake of vitellogenin directly (Campbell, 1978). Oocyte maturation may be affected indirectly by gonadotropin causing the release of interrenal steroids (Goswami *et al*., 1974) or ovarian folliculoid steroids (Jalabert, 1976). The corticosteroids in rainbow trout, northern pike and goldfish appear to function by increasing the follicular sensitivity to gonadotropin (Jalabert, 1976). Jalabert also showed that oocyte maturation by gonadotropin could be dissociated from ovulation. He further showed that the ovulation process could be mediated by prostaglandin and catecholamines. Gonadotropin is also capable of stimulating complete development of the testis in fish (see Billard *et al*., 1982). It is not well established whether this is a direct action or a stimulation via the androgenic steroids. Gonadotropin can elicit the release of androgens and cause the early stages of spermatogenesis. The androgens themselves appear to become more important for the later stages of spermatozoa formation or spermiation. Other factors such as corticosteroids may also be involved with spermiation.

The gonadal activity is activated during a spawning migration by an interaction of the fish with environmental parameters that stimulate the hypothalamus and pituitary gland to produce and secrete their controlling factors. The activity of the gonadal tissue and the changes in circulating androgenic and estrogenic steroids during the migration of different fish species is well reviewed by Woodhead (1975).

Prolactin has various actions in higher vertebrates with respect to reproductive endocrinology and experiments have also been conducted with fish implicating this hormone with various diverse aspects of reproduction. Thus, this hormone has been found to synergize with other hormones such as testosterone to maintain the seminiferous tubules of the Indian catfish (Sundararaj and Goswami, 1965) and the goby, *Gillichthys mirabilis* (de Vlaming and Sundararaj, 1972). Prolactin has also been shown to cause gonadal regression in a number of female sexually mature birds. Such an action may also occur in the Indian catfish (Sundararaj and Keishavanath, 1976) and the golden shiner, *Notemigonus crysoleucas* (de Vlaming and Vodicnik, 1977) although an opposite effect has been observed in the cichlid, *Aequidens pulcher* (Blüm and Weber, 1968) since ovine prolactin injections stimulated certain steroid dehydrogenase activities. Prolactin has already been discussed in this chapter with respect to osmoregulation. One of the functions in this regard was to change the permeability of the skin by the production of mucus. This mucus-producing response to prolactin is exaggerated in the discus fish, *Symphysodon* sp., where integumentary mucus is produced in large enough quantities to be fed to its young. The male sea horse, *Hippocampus* sp., possesses a ventral pouch for the incubation of eggs and prolactin is known to cause a proliferation of the epithelial lining of this pouch in intact as well as hypophysectomized male sea horses (Boisseau, 1969). Prolactin treatment in these fish also stimulates the interrenal cells so that the maintenance of the brood pouch may be indirectly controlled by adrenocorticosteroids.

Prolactin also affects fish reproductive processes through its actions on parental behavior. The incidence of broody behavior in the jewel fish, *Hemicromus* sp., is enhanced following treatment with prolactin (Noble *et al.*, 1938). Several cichlid fishes increase the available oxygen to their developing young by fanning behavior which is increased by the action of prolactin (Blüm and Fielder, 1965; Blüm, 1968). This behavior has also been shown to occur in male three-spined sticklebacks but is stimulated only at low doses of prolactin. It is inhibited at high doses (Molenda and Fiedler, 1971). Bubble nests are built by male paradise fish, *Macropadus* sp., for egg deposition by the female. Prolactin assists in this behavior by increasing the secretion of buccal mucus which aids in the production of bubbles (Machemer and Fiedler, 1965).

The production of urotensin II from the fish caudal neurosecretory center may also be involved with reproductive processes. Urotensin II has a spasmogenic action on a variety of smooth muscle preparations in fish (Berlind, 1973; Lederis, 1977). Although the physiological significance of this spasmogenic action is not known, one possibility is a role in reproduction. Urotensin II causes contraction of the sperm duct in the goby, *Gillichthys mirabilis* (Berlind, 1973) and the oviduct in the guppy, *Lebistes reticulatus* (Chan and Bern, 1976). Several studies have shown that the content of urotensin II in the urophysis varies with reproductive condition, being higher in pre-spawning fish and falling after spawning (Berlind, 1973; Lederis, 1973).

Sensory Physiology

A migrating fish requires a sophisticated sensory repertoire in order to recognize various features from its environment along its own particular migratory route. Chapter 3 dealt with various mechanisms that fish use to orient towards some predetermined goal. These orientation mechanisms are based on an appropriate sensory input. In some cases however the sensory organ or receptor is not as yet known. For example, there is an accumulation of information indicating that some fish species make use of electromagnetic orientation mechanisms but the basis for the required sensory input is open to speculation. The physiological basis for any sensory mechanism need not be static throughout the life cycle of a fish. At times such as migration, specific sensory mechanisms might be required to be activated or existing mechanisms be altered in some fashion such as to make the sensory input more acute or meaningful for migration. In some instances we now know that sensory mechanisms are indeed altered for appropriate responses at different stages during a fish migration.

Olfaction

As discussed earlier, there is considerable evidence that many migrating fish species use olfactory cues to orient towards their home site (see Harden-Jones, 1968; Kleerekoper, 1969; Hasler, 1971). Studies with Pacific salmon further indicate that these fish return to spawning areas where the individual fish itself was reared, not necessarily the place where the parents of the fish returned (see Hara, 1970). The period in time that the young fish are exposed to an olfactory cue appears to be a critical factor.

Electroencephalgram (EEG) recordings from the olfactory bulb have been used by a number of investigators to elucidate olfactory responses of migrating fish (see Bodynick, 1975). The specificity of the EEG response to home stream waters has been investigated as well as changes in the EEG pattern during different phases of fish migrations. Other studies have been conducted using the EEG response to elucidate physiological alterations of olfactory perception, especially as modified by endocrine factors. Although high EEG activity does not necessarily indicate that the fish is discriminating between odors or using olfactory cues for orientation, such patterns do suggest certain changes that are possibly relevant to altered physiological states for olfactory perception. Thus, it is interesting to note that the spontaneous electrical activity in the olfactory bulbs of migrating Pacific salmon was much higher than in non-migratory rainbow trout or goldfish (Hara *et al.*, 1965). Relatively high activity was also found in young salmon during the time of imprinting. Transplantation experiments of young salmon also indicate that imprinting to specific odors can occur in young fish at particular stages in their early life history (Hara, 1970; Scholz *et al.*, 1975).

Imprinting to odors by young fish implies some physiological alteration that can be 'recalled' at an appropriate time to affect a successful completion of a migration. Rappaport and Daginawala (1968) have conducted some interesting experiments with the marine catfish, *Gallichthys felis*, to ascertain a mechanism of memory establishment. They showed that the synthetic odorant, morpholine induced an increase in brain nuclear RNA and a change in base ratios. Later studies by Oshima *et al.* (1969) on homing salmon showed that administration of drugs that block events from transcription to translation markedly inhibited olfactory bulbar discrimination of home water. It thus appears that an olfactory memory in migrating fish for homing may depend on continued metabolism of RNA and subsequent protein synthesis. It is worth noting that memory may possibly become permanent by the secretion of large amounts of melanotropin and corticotropin in young fish (see Meier and Fivizzani, 1980).

Patterns of EEG activity in the olfactory bulb and characteristics of single potentials appear to be influenced by endocrine factors. Oshima and Gorbman (1968) thus found that administration of estradiol to goldfish had a facilitory effect on the response of the olfactory bulb to chemical stimulation. Testosterone also increased the discharges from the olfactory bulb whereas progesterone had a depressant effect. Hara (1967) found that gonadal steroids also affected the individual potentials in the olfactory bulb of goldfish stimulated chemically as well as electrically. The steroids augmented the potentials by changes in amplitude, phase and after-potentials. Thus, increased gonadal activity of fish during a spawning migration may well affect olfactory orientation mechanisms directly or via specific receptive centers in the brain.

Oshima and Gorbman (1966a, 1966b) conducted a series of experiments that implicated the thyroid tissue in olfactory perception. The telencephalon normally augments the olfactory response and these workers found that this reaction could be inhibited in goldfish by thyroxine treatment. Thyroxine was also found to reduce the electrical discharge of the olfactory bulb.

Vision

The sense of vision is important to migrating fish for schooling behavior, home site recognition, food acquisition and possibly landmark recognition during movements within lakes and streams. Therefore, any physiological alterations to visual perception or acuity at appropriate times during migration might enhance visual orientation mechanisms or improve the chances for a fish to acquire food during migration or for premigratory fattening. There is evidence that the visual physiology of some species alters during migration and that different endocrine factors are mediators for various visual changes.

Fish possess different visual pigments to give optimal photosensitivity to the quality of the light that the fish encounters in its particular environment. Thus, the retina of freshwater fish contain porphyropsins as the main visual pigments whereas rhodopsins and chrysopsins are mainly found in coastal and deep water marine fishes respectively. Diadromous species of fish that migrate from one salinity environment to another may thus require changes in visual pigments to optimize their visual sensitivity. Carlisle and Denton (1959) have observed a change in retinal pigments from the porphyropsin to the chrysopsin type in Atlantic eels as they prepare to migrate to the ocean. On the other hand, Beatty (1966) observed that several adult migratory Pacific salmon species displayed an indication of an increased porphyropsin synthesis as the fish migrated into fresh water. These changes were probably not due to changes in diet or salinity. During smoltification in salmonids the visual pigments also change to pre-adapt the fish to the marine environment (Bridges and Delisle, 1974). Also, during smoltification the thyroid gland becomes active which suggests an involvement of thyroid hormones with visual pigment alterations. A number of experiments have now been conducted demonstrating that thyroxine is responsible for converting porphyropsin to rhodopsin (Wilt, 1959; Naito and Wilt, 1962; Munz and Swanson, 1965).

Fish not only exhibit changes in visual pigments but also display electrophysiological alterations associated with visual functions. Hara *et al.* (1965, 1966) have shown that thyroxine affects the optically evoked potentials in the optic tectum in goldfish. They found that following a single light flash, thyroxine shortened the latent period of the response, shortened the time taken to reach the maximum response and increased the amplitude of the positive potential, whereas the thyroid inhibitor, thiourea, produced opposite effects. Thyroxine treatment shortened recovery time when two light flashes were employed. Thyroxine also increased the amplitude of the response following multiple light flashes. These authors suggested that thyroxine may have caused a morphological or metabolic change in specific parts of the visual system and not just simply overall neuronal activation.

Gonadal steroids have also been shown to affect visual sensory perception (Hoar *et al.*, 1952; Hara *et al.*, 1965). Nervous excitability can be enhanced and sensitivity to odors is increased by the hormones. Since gonadal steroids are produced during spawning migrations, these effects of increasing visual acuity would be useful for upstream migrations and recognition of the home site or spawning ground.

Salinity Preference

For a fish to prefer and seek out a particular environment with an appropriate salinity there is a need for a suitable behavioral response. Behavioral alterations are likely to be brought about initially by changing physiological conditions within the fish. Although the end result is manifested behaviorally, salinity preference is based upon sensory perception. Some fish show a definite response to choose a particular salinity environment during specific times in their life cycle and in order to recognize when they are in water of the desired salinity, they need to possess some type of sensory input. Whether the sensory input comes from osmoreceptors in the brain or elsewhere in the fish, or whether there is some other mechanism remains to be determined. However, there is information available indicative of various external and internal parameters causing a change in salinity preference.

Migratory fish that are diadromous require the ability to adapt to waters of differing salinities. Just prior to entry of some fish into a new salinity environment there is a change in salinity preference towards the ionic concentration of the water into which the migrant will soon enter. In the northern part of their range adult three-spined sticklebacks migrate into fresh water to spawn and just prior to leaving sea water these fish display a preference for fresh water. Contrarily, Pacific salmon migrate to the ocean as young fish and exhibit a strong preference for sea water at the time the freshwater migration begins. Environmental factors initiating such changes in preference to different salinities appear to be photoperiod and temperature (see Meier and Fivizzani, 1980). Increasing photoperiods thus seem to stimulate a change in salinity preference for various young Pacific salmon (Baggerman, 1959; McInerney, 1964; Otto and

McInerney, 1970), the gulf killifish, *Fundulus grandis* (Spieler *et al.*, 1976b) and the three-spined stickleback (Baggerman, 1959). Gunter (1967) has attributed the movement of a number of estuarine species of fish into fresh water to increasing temperatures.

At the time of migrations of the above-mentioned fish there is an increase in thyroid tissue activity and thus the possibility that thyroid hormones might be the cause of changing salinity preferences. Baggerman (1957, 1960, 1963) has conducted numerous experiments with Pacific salmon and three-spined sticklebacks indicating a role of thyroxine in salinity preference. However, it is difficult to assess the role of thyroxine from these experiments since it causes a preference for fresh water in the sticklebacks and for sea water in the salmon. McInerney (1964) conducted a comparative investigation on five species of Pacific salmon with respect to salinity preference and found that differences between the species correlated with the different times that each species actually migrates. Thus, the change in salinity preference may be the stimulus to initiate migratory behavior in these species. Fivizzani and Meier (1978, cited in Meier and Fivizzani, 1980) have provided further evidence that prolactin may also cause a change in salinity preference in the gulf killifish but only if administered in conjunction with cortisol. This action may be a further extension of the role of thyroxine, but this of course awaits further investigation.

Physiological Changes Related to Behavior

The entire phenomenon of fish migration is basically a behavioral response to various environmental parameters in order to enhance food acquisition or ensure reproductive success. This behavior is initiated by various factors and includes activities such as swimming, feeding and orientation. Whatever the particular behavior might be, it has a basis in some physiological alteration. The behavioral modifications required for a successful migration may be brought about physiologically in a number of different ways. Behavioral manifestations for migration may be due to functional changes in the central nervous system and in addition may be initiated by alterations in the physiology or morphology of peripheral structures. Thus, physiological modifications may take place in peripheral sensory receptors that cause a stimulatory or inhibitory effect on particular sensory inputs. The central nervous system may elicit specific behavioral patterns due to altered metabolic processes in brain tissue. For example, thyroxine treatment in lungfish enhances cholinesterase activity in the brain whereas thyroidectomy reduces the activity of this enzyme (Dupe and Godet, 1969). Cholinesterase is an enzyme known to have effects on nerve activity. The central nervous system in fish may also modify or elicit particular behavioral sequences by the augmentation of the sensitivity of specific brain centers. An example of such specialized brain areas which evoke particular behavior patterns comes from electrical stimulation or brain lesioning studies in fish that have demonstrated the lateral hypothalamus as a specific brain center controlling feeding (see Peter, 1979).

Initiation of Migration

Changes in sensory input from peripheral receptors in association with certain brain centers are likely to be responsible for initiating migratory behavior. Environmental parameters which at one time might be ineffectual become important through sensory input which results in migratory behavior. Since migration is a seasonal behavioral response, fish may change their internal or endogenous physiological rhythms to react at appropriate times to some proximate environmental stimulus. These endogenous rhythms may in turn be set by annually changing environmental parameters. Such external stimuli as photoperiod and temperature may thus set annual physiological rhythms and therefore ultimately control the elaboration of fish behavioral patterns (see Meier and Fivizzani, 1980). The proximate environmental factors that initiate migratory behavior have been discussed in Chapter 3, but here it is of interest to discuss the internal environment of the fish that allows it to respond accordingly to these external stimuli.

It is tempting to speculate that specific levels of circulating gonadal hormones are required to initiate migrating activity. On the surface this would seem to be the case since there are many fish migrations that take place when gonadal tissue is developing and the fish are moving to spawning areas. Also, there are a few studies indicating that gonadal steroids can increase locomotory activity in a number of different fish (see Woodhead, 1975). However, it is difficult to reconcile these findings or associations with the fact that many fish species undertake long migrations when their gonads are at an early stage of differentiation or when they are sexually immature. Males and females of various fish species migrate at the same time but gonadal maturation often occurs in one sex before the other. However, it may be argued that concentrations of steroids in the blood may act differently between males and females and that there are possibly vast differences in androgens and estrogens with respect to migration initiation. Within a given species there are great differences in time when the gonads develop in relation to spawning migrations. Within the Pacific salmon, for example, there are some races that exhibit gonadal maturation before migration whereas others reach the spawning grounds before gonadal development starts. Obviously, more research is required in this area to elucidate the role of gonadal steroids in initiating a migration. However, the situation may seem so variable at this time because some fish may depend on gonadal steroids whereas other species may not, and some species may require quite different types of steroids or concentrations to reach a threshold required to begin a migration.

An active thyroid tissue has been described for many fish preparing to migrate. The thyroid cells (Hoar, 1939) and the pituitary TSH cells (Olivereau, 1954) both increase in numbers during the onset of seaward migration in salmonids. However, this is at a time when various changes related to smoltification are occurring so that it is difficult to assess whether the thyroid activity is also related to the onset of migration. Thyroid activity is also associated with the upstream (Sklower, 1930) and downstream migration (Bernadi, 1948) of European eels. The previously discussed role of thyroid hormones in salinity preference may also be an indication of thyroid involvement in starting migratory behavior. Hoar *et al.* (1952, 1955) demonstrated that thyroxine treatment stimulated swimming activity in certain Pacific salmon species as well as in goldfish. Oxygen consumption is also increased in fish ready to migrate and may be related to thyroid activity (Baraduc and Fontaine, 1955). These studies may be indicative of a pituitary-thyroid axis stimulation to initiate migratory behavior, or may be a reflection of an action of the thyroid on other factors or mechanisms since thyroidectomy in steelhead trout and Atlantic salmon still leads to downstream migration (see Hoar, 1965; Norris, 1966).

The interrenal cells of many fish species become active during or prior to migration. The circulating levels of corticosteroids become elevated and interrenal cells hypertrophy (see Woodhead, 1975). These changes occur both in young fish as well as in spawning migrations. The problem still remains, however, as to the role of these steroids in initiating migration as increased interrenal cell activity may only be a reflection of other actions such as those previously discussed for osmoregulation and catabolic reactions.

There is some interesting work implicating prolactin as a factor responsible for migration initiation in certain vertebrates. The amphibian newt, *Notophthalmus viridescens*, migrates from a terrestrial environment back to water where it metamorphoses into an aquatic adult. This migration or 'water drive' has been demonstrated to be a result of the action of prolactin. Water drive is also apparent in other amphibians and in some this behavior is also linked to the action of prolactin (see Clarke and Bern, 1980). Meier (1969) has also demonstrated a behavioral change in response to prolactin in the two avian species, the white-crowned sparrow, *Zonotrichia leucophrys,* and the white-throated sparrow, *Z. albicollis.* These two species migrate at night and if caged they show increased nocturnal locomotory activity or restlessness during the migratory period. Prolactin injections can thus induce this restlessness at periods prior to when this behavior normally occurs. Prolactin may also be involved with initiating migratory activity in fish. The marine form of the three-spined stickleback, *Gasterosteus aculeatus trachurus*, spends most of the year in sea water but migrates into fresh water in the spring to reproduce. The survival of this fish in fresh water requires prolactin (Lam and Hoar, 1967; Lam, 1969, 1972; Lam and Leatherland, 1969, 1970). The prolactin cells change in their activity during the year and become most active during the spawning migration although there is increased activity well before the fish actually enter fresh water (Leatherland, 1970). Prolactin cells have also been shown to become active in the chum salmon prior to the entry of the fish into fresh water (Nagahama, 1973). Adult migratory sockeye salmon also possess prolactin cells that contain granules and appear active when the fish are still in sea water (McKeown and van Overbeeke,

1969; Cook and van Overbeeke, 1969). A later study on the same species by McKeown and van Overbeeke (1972) assessed pituitary and serum prolactin levels of fish at sea, at various locations near entry into fresh water, in the river and on the spawning grounds. Interestingly, the fish released prolactin at the time of entry into fresh water but also exhibited high levels of prolactin in both the pituitary and blood while the fish were still 360 km from the mouth of the river. There was also an indication that prolactin was released from the pituitary before the fish encountered fresh water although the sea water was of a lower salinity when this occurred. As with other hormones, there are no definitive studies with fish that would indicate that prolactin is solely or in part responsible for initiating a migration. These previous studies may thus only indicate a pre-adaptation role of prolactin for subsequent survival of the fish in fresh water.

Swimming Activity

After migration has been initiated it will come to a completion or the goal will be reached only if appropriate activity patterns are exhibited and controlled. The relevant patterns of behavior include such aspects as orientation, reproduction, interspecific relations, intraspecific associations, and swimming. Once oriented in the direction towards the home site or goal, swimming activity itself may depend also on a number of variables such as food availability, amounts of body reserves, and external biotic as well as abiotic modulating factors. There is thus a need for a physiological control over swimming activity due to variations in some of these prevailing modifiers.

Chum and coho salmon fry respond to thyroxine treatment by increasing their swimming speeds and reverse this response when treated with an anti-thyroid drug (Hoar *et al.*, 1952, 1955). However, Godwin *et al.* (1974) showed the opposite effect of thyroxine when administered to yearling Atlantic salmon. These authors point out that the observed reduction in swimming activity may be a differential response of Atlantic salmon or may be due to the low doses they injected. However, the original experiments administered thyroxine via addition to the aquaria water and thus the actual dose the fish received is not known. Hoar *et al.* (1955) also showed that the adult goldfish increased its swimming activity when immersed in a thyroxine solution. Similarly, swimming activity following thyroxine treatment is also enhanced in brown trout larvae (Woodhead, 1966), the guppy, *Poecillia reticulata* (Sage, 1968) and the cod (Woodhead, 1970). Treatment of castrated three-spined sticklebacks (Baggerman, 1962) or immature European eels (Gineste, 1955), with thyrotropin, which presumably stimulates thyroid activity, also stimulates swimming activity. There are also a number of studies that correlate activity patterns with increased thyroid hormone secretory activity (see Woodhead, 1975). Thus, if certain species of fish are forced to swim or are naturally engaged in more active swimming, the thyroid activity is elevated. Comparisons of different fish species also indicate that the more active species have higher thyroid activities. However, the cause or result relationships are not known. On the other hand, these findings may be a result of thyroid hormones or contrarily, the increased thyroid activity may only be a reflection of increased metabolic demands which in turn brought about an increase in thyroid hormone secretion. The experiments by Hoar *et al.* (1955) are of interest with respect to how thyroid hormones may have an action on locomotory activity. These hormones may play such a role by affecting nervous excitability. They could stimulate goldfish to swim a standard distance by applying an electrical stimulus. Following thyroxine-treatment, this response could be obtained with a weaker stimulus. Also, the reaction time for salmon to jump from one experimental tank to another was reduced after treatment with thyroxine.

Since gonadal maturation occurs in many fish during migration, it is not surprising that gonadal steroids can also increase locomotory activity in a number of teleosts. These hormones, like thyroid secretions, are also involved in metabolic processes which may be indicative also of an activity related function. Such a finding was reported over a half century ago by Stanley and Tescher (1931) who recorded increased swimming speeds in goldfish following oral administration of testicular tissue. Hoar *et al.* (1952, 1955) confirmed this result in goldfish by exposing them to water containing either testosterone or a synthetic estrogen. Likewise, male sticklebacks are more active during the breeding season whereas castration reduced swimming activity (Baggerman, 1963, 1966). Replacement therapy with testosterone in the castrated fish increased swimming activity once again. Many fish species are more active during gonadal recrudescence but the cause or result relationships of the gonadal steroids and swimming activity are of course unknown (see Woodhead, 1975). As with thyroxine, Hoar *et al.* (1952, 1955) found that the observed increase in swimming activity following gonadal steroid treatment may be due to an effect on nervous excitability since these steroids lower thresholds for jumping activity in salmon and electrically-stimulated swimming in goldfish.

Northcote and Kelso (1981) have conducted some interesting experiments to ascertain whether rainbow trout populations above impassable waterfalls inherit positive rheotactic behavior patterns in order to maintain such populations. By testing appropriate crosses of parent fish, above-waterfall populations do seem to inherit specific rheotactic behavior patterns. With respect to swimming activity it is of interest to note that these fish displayed distinct lactate dehydrogenase (LDH) phenotypes. These authors, as well as Klar *et al.* (1979), suggest that physiological alterations related to changes in swimming ability might be due to different LDH types. Thus, the possibility exists that observed physiological differences between fish, such as swimming activity, might be due to a genetically controlled enzyme system.

Feeding Activity

Fish, like other organisms, require appropriate morphological, physiological and behavioral mechanisms for acquiring and utilizing food materials. These mechanisms become more important during migration and periods of premigratory fattening. Some species feed before migration is initiated, deposit the food materials as appropriate energy stores and then mobilize these materials as required during migration. On the other hand, there are species that have migration routes where food acquisition is possible along the way. Either option, however, requires the appropriate physiological controls over feeding activity.

Feeding behavior and physiological mechanisms regulating it in teleosts have been well reviewed recently by Peter (1979). He discusses the fact that growth rates are limited by food availability as well as a variety of other factors such as: intraspecific social interactions, interspecific social interactions, size, migration, spawning, temperature, season, salinity, oxygen concentrations, disease and toxicants. Irrespective of these modulating factors fish appear to control food intake to regulate body weight to some 'set-point'. Evidence for this type of regulation comes from experiments such as those of Rozing and Mayer (1961) whereby goldfish were found to compensate for the size and nutritive value of food pellets. As the size of the pellets increased, the number ingested decreased. As the nutritive value of the food increased the quantity ingested again decreased. Johnson (1966) found that both food intake and growth in the pike were regulated on an annual basis irrespective of the amount of food available. These data too would lend credence to the establishment of a 'set-point' by fish. Further evidence comes from food deprivation studies. Various species appear to compensate for periods of starvation by increasing ingestion when food again becomes available (see Peter, 1979). Tyler and Dunn (1976) found that the winter flounder, *Pseudopleuronectes americanus*, appeared to increase the meal size when the meal frequency decreased.

The natural environment does not provide food *ad libitum* at all times of the year. When food is in short supply individual fish are generally of a smaller size but, in a similar situation, Nakashima and Leggett (1975) found the population size also decreased in yellow perch, *Perca flavescens*. Food availability may thus modify feeding activity and at times of low food intake gonadal recrudescence may not occur which thus would lead to reproductive failure and a decrease in population size. Migratory success with its associated increase in energy demands may likewise be indirectly associated with feeding activity.

Peter (1979) has reviewed studies in fish indicating brain regions involved with food intake. The lateral hypothalamus has some role in controlling fish feeding behavior. It appears that more work is required to elucidate the brain centers responsible for satiation, arousal of feeding and regulation of body weight set-point level. Blood glucose levels influence food intake in mammals but little is known for such a role in fish. Peter *et al.* (1976) have conducted studies in goldfish employing gold thioglucose and concluded that blood glucose levels are not important in the regulation of food intake in fish. Although gold thioglucose caused brain lesions in particular areas of the brain due to the toxic effects of gold, localization was not likely due to glucoreceptors since gold thiomalate had similar effects.

There is even less information available indicating a role for various amino acids or free fatty acids in regulating food intake in fish. However, there is an indication that fattening does occur in various fish species under the effect of different hormones (see Meier and Fivizzani, 1980). Cortisol and thyroxine thus have been shown to elicit a fattening response to prolactin. These hormones are in turn affected by environmental changes such as temperature and photoperiod (see McKeown and Peter, 1976) that at appropriate times of the year can thus induce fattening. Unfortunately, the roles of the hormones that cause fattening are not known with respect to food intake. In fact, these hormones possibly might not even be involved with regulation of food intake but only with fat mobilization, irrespective of food ingestion.

There is now considerable information available concerning endocrine factors responsible for mobilizing energy reserves for body growth and differentiation in fish. There are also data indicating a role for some of these factors in regulating food intake.

Initial work by Pickford (1957) indicated that bovine growth hormone could cause an increase in food intake in hypophysectomized killifish. Subsequent studies demonstrated a similar effect of mammalian growth hormone preparations in Atlantic salmon parr (Komourdjian *et al.*, 1976a) and yearling coho salmon (Higgs *et al.*, 1975, 1976, 1977, 1978). Markert *et al.* (1977) indicated an increased appetite in bovine growth hormone-treated coho salmon and attributed the effect to either the hormone acting directly on appropriate food intake centers in the hypothalamus or inducing metabolite changes which in turn indirectly affect appetite centers in the brain. With respect to growth hormone, the latter possibility seems to be the case since growth hormone levels in goldfish were observed to be increased at a time when food intake and growth were decreased (Peter *et al.*, 1976), Gonadal steroids also appear to affect food intake as methyl-testosterone increases appetite in the goldfish (Yamazaki, 1976) as does diethylstilbestrol in the plaice (Cowey *et al.*, 1973).

Thyroid hormones are also known to effect metabolite alterations in fish that in turn may elicit increased feeding activity. The possibility also exists that the thyroid hormones directly affect specific brain centers. Thyroxine treatment increases food intake in the green sunfish, *Lepomis cyanellus* (Gross *et al.*, 1963) and the coho salmon (Higgs *et al.*, 1977). It will be interesting to follow future research to ascertain the way in which hormones and other factors directly and/or indirectly affect feeding activity. This is from the point of view of basic research as well as for possible practical applications in aquaculture.

6 ECOLOGY AND EVOLUTION

Introduction

The topic of this chapter has been well reviewed recently in other book chapters (see Northcote, 1978; Dingle, 1980). No review on the ecology and evolution of animal migration could be written without due respect and recognition of the excellent and extensive works by Baker (1978, 1982). Those who are interested in the detailed descriptions, evaluations and interpretations of animal migrations in general are referred to these books by Baker.

A number of authors have postulated and evaluated the various reasons why fish migrate. Northcote (1978) has summarized the various strategies for freshwater fish migrations, but the reasons given in his review could easily apply to fish migrations in general, whether it be in fresh water, sea water or movements between the two. Northcote suggests that fish migration has evolved (1) to optimize feeding, (2) to avoid unfavorable conditions, (3) to enhance reproductive success and (4) possibly to promote colonization. Specific fish migrations need not satisfy all these conditions and indeed might be for only one of the above reasons. Also one particular phase of a migration might accomplish more than one purpose.

If a species migrates, it will always move away from, and at some later time return to, the same or a similar spawning area to that in which the individuals were hatched. The spawning area is by necessity also the first step of any migration. However, the greatest changes in biomass usually occur in the feeding and wintering areas and as such markedly affect production which in turn imparts selective pressures on the movement patterns to and from these areas. Migrations, however, are brought about because reproductive habitats are not likely to be the best areas for other activities such as feeding and overwintering. This is especially true for fishes that inhabit temperate and polar regions where there may be pronounced seasonal changes affecting the habitat. Any fish species will move about to optimize the utilization of a habitat's resources during different phases in the life cycle of the fish. These movements may occur from minute to minute or from season to season. Although the movements are of vastly different duration the selective pressures to optimize utilization of resources may be very similar. Not only does a seasonal change in habitat quality select for certain types of migratory behavior, but also, so do the changes that occur in the internal environment within the fish. For example, some fish species return to their natal areas at different ages. Certain individuals thus attain a physiological readiness for reproduction at an earlier age. Those individuals that do not reach the appropriate reproductive physiological state often migrate to a different area such as a feeding area or some habitat to avoid an unfavorable condition.

Likewise, most one-year-old sockeye salmon, *Oncorhynchus nerka*, smolts migrate out of their nursery lakes but a small proportion of smaller smolts remain in the lake for one additional year. These smaller fish are presumably not in an appropriate physiological condition of readiness to migrate.

The possible patterns of movement between different habitats are shown in Figure 6.1. Many migrations are cyclic and bring the fish back to one area

Figure 6.1: Generalized patterns of migration between the three basic habitats utilized by many migratory fishes.

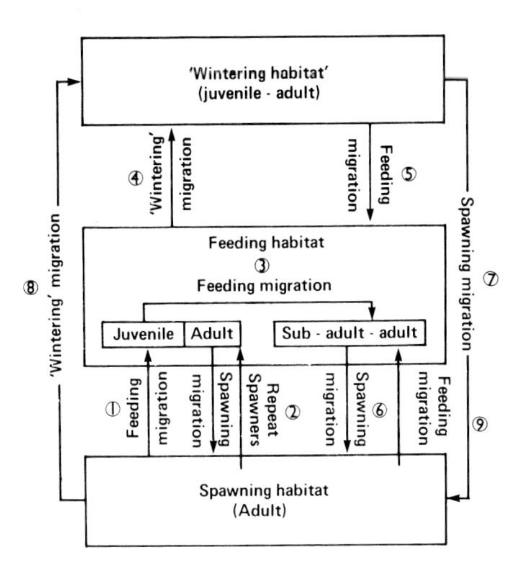

Source: redrawn from Northcote, 1978

frequented earlier. Some habitats are not separate, as for example when young fish use the spawning area as a nursery and feeding area as well. Not all areas will be used by some fish, especially in tropical areas where a habitat to avoid an unfavorable condition brought about by seasonal changes often does not exist. Thus, some fish may move out of a natal area to a feeding area and return to the spawning area as sexually mature adults. Cycles may be repeated a number of times. Fish may move between feeding areas to avoid unfavorable conditions more than once until they reach an appropriate size or physiological state allowing a spawning migration to take place. Also, a number of fish are repeat spawners and complete return migrations a number of times. The second and subsequent migrations of repeat spawners are also not necessarily the same as the first return migration to the spawning grounds. Figure 6.1 indicates spatially separated habitats. However, any one of the different types of areas may have similar habitats but in a different location. Fish may thus move from a feeding area to a wintering area and then return to a feeding area. However, the second feeding area is not necessarily the same one visited before but may have similar resource characteristics. As fish grow and mature they require different food materials and a second feeding area is often required to meet the new requirements of the developing fish. In any given population of fish, most of the individuals might migrate to the next appropriate habitat. However, some of the individuals in that population may move to a different area but still an area that has similar characteristics to the one towards which most of the others migrated. This divergence in the migratory pattern between individuals of the same population may be brought about by a number of different factors and might occur regularly due to divergence in the cues used for orientation. For example, a population of fish moving along with an ocean current could easily have individuals separated from each other if there is a bifurcation in the current. The divergence in migratory patterns may, however, not necessarily be a regular phenomenon and thus may occur only occasionally due to irregular perturbations in the habitat through which the fish are migrating. Temporary and irregular changes might occur in current patterns or a spawning area in a small stream might be eradicated or changed by landslides, forest fires or volcanic activity thus forcing individuals into new areas. These types of diversions may be important in promoting colonization of new areas.

Strategies of Migration

Feeding Optimization

In many instances young fish migrate away from natal areas to highly productive feeding areas. Indeed the fish may spend most of their life in these productive areas, only returning briefly to the less productive spawning areas as sexually mature adults. If feeding is a strategy to increase growth and thus survival and fecundity, fish might be expected to leave spawning areas and migrate to feeding areas as early as possible. Fish generally do migrate to feeding areas early in their life cycle although some species feed in or near the natal site. The fish that delay their migration to a feeding area may do so because they must develop an energy reserve first in order to make the journey or because they may require a development of structures or mechanisms to allow them to adjust to the new environment.

The timing of migration patterns appears to have evolved such that young fish arrive at feeding areas when environmental conditions are optimal for food production. Many oceanodromous species thus deposit their eggs in locations where ocean currents carry the developing pelagic young through feeding areas at appropriate times of the year. The seaward migration of young salmon occurs at different times of the year. Populations in southern latitudes move into the ocean sooner than populations of the same species that migrate from rivers further north. This timing difference coincides with the earlier increased food production in southern waters.

Many young fish hatch in habitats poor in food availability and their first migrations are primarily for feeding. There are many examples of species that move further downstream or move out of streams and feed in lakes. Many anadromous forms move quickly into coastal marine habitats or estuarine areas which serve as nursery feeding grounds. Once the young fish enter a feeding habitat, they often move in large schools feeding from place to place. Such fish often complete migration circuits before reaching maturity and undertaking a migration to the spawning grounds. The sequential exploitation of different parts of the feeding habitat may be a means whereby fish optimize production for a whole population.

Another strategy that fish employ during migration seems to be a short-term and rapid exploitation of a rich food source. River flood-plains or temporarily flooded littoral zones of certain lakes often provide a habitat that is rich in a food of terrestrial origin. These zones are only available for short periods at times of heavy seasonal rain. In addition to food of terrestrial origin, even the eggs and fry of other fish may provide an abundant temporary supply of food. For example, it is not uncommon to observe spawning Pacific salmon being accompanied by Dolly Varden (*Salvelinus malma*), cutthroat trout (*Salmo clarki*), whitefish and sculpins. These predators migrate to the salmon's spawning ground and feed voraciously on eggs. The fry of Pacific salmon often emerge and migrate collectively in large numbers. This again provides the opportunity for predator species to migrate to these areas of fry movement and capitalize on a short-lived but rich food source. Habitats that display temporarily rich food resources often exhibit other characteristics which are of an adverse nature. Some shallow stream or estuary environments may provide good food resources at certain times of the year but also often show drastic daily changes in temperature or oxygen as well as increased risks of predation. These areas might not be permanently inhabited because of the predation risk and the temporary nature of food availability.

Fish migrate to feeding areas presumably to enhance food intake. Increased feeding would in turn cause individual fish to increase growth rates, fecundity and/or survival. These increases may in turn increase production of a species. There are many examples of species which have both migratory and non-migratory populations (see Northcote, 1978). The migratory populations often move to rich feeding areas and subsequently return to similar spawning areas as the non-migratory forms. The migratory individuals are usually much larger and more fecund. The fish that migrate to better feeding areas are sometimes more abundant as well which may be a reflection of growth and fecundity.

Female fish often reach larger sizes than males and must also divert more energy reserves into gonadal development than do the males. This being the case one would expect that the female fish would migrate more readily or for longer periods of time than males to productive feeding areas. Northcote (1978) has reviewed the literature and a number of species exhibit differences between male and female migration patterns. A number of anadromous species have populations where the female migrates to the sea but the male remains in fresh water and although much smaller still reproduces with the sea-going female

(Österdahl, 1969; McPhail and Lindsey, 1970; Maksimov, 1972). Many populations of Pacific salmon display migrations to the sea of both male and female fish. However, in many instances a fairly significant proportion of the males return to fresh water to breed as precocious fish or 'jacks'. This precocious behavior may be a reflection of a lower feeding requirements in males, but may have evolutionary significance for other reasons.

There is evidence that feeding migrations take place more frequently in temperate and polar regions than in tropical or subtropical areas. This might be directly related to the food availability since primary productivity is much higher in temperate than tropical regions. Seasonal changes in productivity are greater the further one travels from the Equator. This cyclical nature of primary productivity in temperate regions adds a degree of predictability to levels of food availability. The development of a regular pattern of feeding migrations would thus be much more likely in temperate and polar regions. A number of feeding migrations that do exist in tropical waters appear to be restricted to estuarine areas. Food productivity is usually higher in such areas and seasonal changes in freshwater discharge from rivers might add a degree of predictability that could lead to selection for feeding migrations to such areas.

Although patterns of fish migration show that fish move to areas of food, the stimuli for such movements are less well understood. There is circumstantial evidence as outlined above that migration is related to food supply and that fish productivity is enhanced in these areas of higher food availability. Experimental enrichment of food resources in the natural environment under conditions of aquaculture and mariculture has led to increases in fish growth and population densities. Field observations and measurements by Dill *et al.* (1981) have indicated that the feeding territory of juvenile coho salmon, *Oncorhynchus kisutch*, decreases in size as the density of the benthic food on the territory increases. The reduction in territory size was not caused indirectly by attraction of non-territorial fish to areas where food was abundant. Laboratory studies indicated that the distance from which a resident coho attacked an approaching intruder increased with hunger. These authors suggested that the fish therefore, appear to possess an appropriate behavioral mechanism to adjust territory size to local food abundance. As food becomes less abundant the territory size will thus increase and a given habitat will therefore contain fewer individuals. Presumably then, the fish not maintaining a feeding territory are forced to migrate away by the more aggressive dominant resident fish. The degree of hunger was shown to enhance aggressive behavior in juvenile coho salmon and presumably this lack of food forces some individuals to migrate to better feeding areas. Hartman *et al.* (1982) have added further support to the hypothesis that lack of food due to intraspecific interaction leads to downstream migration of coho salmon fry. They investigated the seaward migration of fry over a ten-year period from a coastal stream in British Columbia and related the movement of the fish to various environmental parameters. In most years the migrant fish are smaller than the remaining resident individuals. Hartman *et al.* suggest that the smaller size of the migrants was a consequence of their subordinate status and that there is, therefore, a behavioral population regulatory mechanism. This mechanism is thought to be brought about by food availability and leads to a displacement downstream and thus seaward migration to better feeding areas.

Avoidance of Unfavorable Conditions

Unfavorable chemical, physical and biological conditions may occur in an environment at different times. Such fluctuations occur because of seasonal changes but may also arise due to daily changes or other biotically controlled cycles. Many marine and freshwater fishes move to so-called wintering areas because the seasonal temperatures in the feeding areas become too low. Many species of fish in tropical freshwater environments also migrate in dry seasons to avoid being trapped in shallows that might dry up, overheat, provide an inadequate oxygen supply, lead to predator exposure, etc.

Seasonal migrations such as the above occur largely due to climatic changes and are thus regulated on an annual basis. Various parameters also change within a habitat on a daily basis and migrations can thus also be timed according to the change in the daily light-dark cycle. A number of fish species thus perform daily vertical migrations. In some respects these fish movements might be considered as feeding migrations since the food resource also moves virtically according to a daily rhythm. However, a fish such as herring, *Clupea harengus*, that feeds on various plankton species, moves deeper than the plankton during the day and moves up into the plankton zone only at night. The exact reason that vertical fish migrations occur is still hypothetical but the fish are probably moving to avoid some unfavorable condition. The herring may thus remain at greater depths during the day to avoid being seen by predators in the brighter shallower waters. They might also be avoiding the warmer temperatures so they can convert more energy into body growth when their basic metabolic rate is lower in colder waters.

Other fish such as the bluegill sunfish, *Lepomis macrochirus*, migrate short horizontal distances offshore where there are dense levels of vegetation (Goodyear and Bennett, 1979). These migrations appear to be in response to predator pressure.

In order to avoid an unfavorable habitat for some time during the day, year or life of the fish, some species move to new areas that are not particularly advantageous with respect to feeding but can still be desirable areas because of a better climate or avoidance of predators. If the new area does not meet the nutritional requirements and the fish needs to stay in the area for a prolonged period of time to avoid unfavorable situations elsewhere, an adequate food supply must be carried to the 'avoidance' area. Thus, many overwintering populations of fish must feed enough in the summer or warmer months in the feeding areas in order to store reserves that can later be called upon when required. In overwintering areas the conditions are such that fish are not usually active and thus have less energy requirements. Nevertheless, bioenergetic pre-adaptation is still often required for these 'avoidance' areas.

Optimization of Reproductive Success

The movements of fish to and from feeding areas to avoid some unfavorable condition permits individuals not only to survive but also to produce somatic and gonadal growth. Populations of any species of fish would not survive from generation to generation, however, unless an appropriate spawning habitat was found. Such habitats are often far removed from feeding areas or areas used to avoid unfavorable conditions and are often unsuitable for purposes other than reproduction. Developing eggs, larvae and young often require vastly different conditions to those required by adults. Many spawning adults thus migrate away from spawning areas shortly after reproduction or in some cases such as Pacific salmon, die after spawning. Death of spawned adults is not necessarily a reflection of the unsuitable nature of the spawning habitat but may be an indication of the bioenergetic cost of the spawning migration. The post-spawning death of adult fish may also provide a mechanism whereby the spawning or subsequent nursery areas of the young fish are enriched. Many fish spawn in relatively nutrient-poor environments that might be enhanced by the dead adults adding nutrients to lower tropic levels. By the time the fish hatch appropriate food sources would become available on or near the spawning area.

Fish are often dispersed widely over large areas while feeding but upon arrival at a spawning area, densities can increase significantly. The congregation of individual fish into specific spawning grounds may be a reflection of only limited appropriate reproductive areas being available, but it does promote reproductive success in many species. The arrival of spawning adults at the same time at spawning grounds coupled with specific behavior patterns, spreads the fish out over suitable spawning areas, thus optimally utilizing the habitat chosen for reproduction. Precise timing of spawning migrations also ensures the simultaneous presence of sufficient numbers of both sexes, in cases where the sexes migrate separately. The main stimulus thus appears to be for fish to migrate to the spawning area after which courtship behavior, pair bonds and reproduction may take place. Since many individual fish migrate to reproductive areas at the same time, a rich food source becomes available and is exploited by a number of fish predators which causes the loss of some individuals. Thus, if the primary goal is the orderly arrival of sexually mature adults to the spawning area, secondary mating will be enhanced as many fish arrive at similar times.

The timing of the spawning migrations becomes important for a number of reasons. Synchronized arrival of a population at a specific location can ensure reproductive success of that group. If the spawning area is spacially separated from other areas, populations may become reproductively isolated even though they might belong to the same species. Such isolation may be important to ensure the inheritance of specific orientation mechanisms, physiological adaptations and behavioral manifestations that result in specific migration patterns for the particular fish population. Such group characteristics would be most important to ensure that sufficient numbers of spawning adults arrive synchronously. Reproductive isolation may also occur between different fish populations within the same species even though the different populations utilize the same spawning grounds. Instead of the fish populations being spatially separated the spawning migrations occur such that the different populations arrive at the spawning areas at different times and are thus temporally separated. The temporal separation of spawning adults of different species is also important in order to avoid over-utilization of specific areas during confined periods of time.

Each species of fish has detailed requirements of a habitat to ensure successful reproduction and survival of its young. Migration patterns thus seem to have been evolved to ensure that spawning adults reach such areas. However, if all fish homed precisely to their own natal areas they may become too fixed and eventually suffer reproductive disadvantages (Harden-Jones, 1968). Many spawning areas change annually. In order to be successful, migrating fish must be flexible in areas chosen for spawning. This statement seems to be contrary to previous discussions but it is only a matter of degree, so that a particular fish population does not become too specific in its migration traits and thus fragile in the process.

Optimization of Colonization

Over long periods of time habitats may change and in so doing may affect migration patterns of fish. Habitats may change such that new areas become available or former areas may again be recolonized due to the removal of some temporary unfavorable condition or blockage of a migration route. Glaciation and deglaciation in the northern hemisphere no doubt has led to changes in salmonid fish migration patterns. Indeed glaciation may have brought about the movement of many salmonids into the marine environment. The benefits gained from rich feeding areas at sea might explain the retention of anadromous behavior long after deglaciation has taken place. Colonization of new habitats is still proceeding for some Pacific salmon species that have been introduced into Russia and eastern Canada.

Mountains formed by folding, upwarping, fault-blocks or volcanic activity and mass movement of surface material such as landslides, mudflows, earthflows, talus, creep, solifluction and rock glaciation have caused considerable changes in recent geological times. These geological changes have often affected drainage patterns and the alterations have often been abrupt and consequential. When migratory routes have been cut off or altered drastically, species have survived by colonizing new areas. The establishment of homing to these new areas is an intriguing problem for both geneticists and physiologists. Severe or long term seasonal fluctuations in the environment force migrant fish into new habitats.

Lifetime Tracks

Baker (1978, 1982) has reviewed and added to our understanding of the evolutionary ecology of animal migrations in ways that far exceed my attempts here. However, in the interests of completeness, I have included here a summary of various aspects of Baker's works and ideas. The lifetime track of animals is a central theme to Baker's writings. He states that

> the most successful individuals in terms of the number and quality of offspring they produce [i.e. reproductive success], will be those with a path that exploits the environment to the full, minimising costs, maximising benefits, always seeking the best trade-off between the two. The path that an animal produces is its lifetime track, the outward manifestation of the individual's solution to spatial and temporal problems, the playing off of inherited predispositions and acquired experience against the environmental backcloth, running the gauntlet of natural selection for yet another generation.

Most animals, including fish, include in their lifetime track a daily return movement to a place of sleep or rest. From these places of rest, the animal moves out each day for a variety of purposes such as feeding, territorial defence, etc. Intertidal fish may return daily or twice daily, to the same tide pool every time the tide ebbs. In some instances the point to which a fish returns is very precise and the degree of return is thus high. Sometimes the daily movement cycle takes the animal to a different but similar location for rest or some other purpose but the pattern is still similar. Each of these movement patterns, however, adds to the lifetime track. The track of an animal covers many different points and when all points are put together they form the range of the animal. The range may be considered on a daily, monthly, annual and lifetime basis. If animals repeat daily movements to and from the same places then the degree of overlap is high, but if there is minimal overlap the animals might be considered as nomadic. Some nomadic types of movements may display a minimal degree of overlap for a daily range but the annual range may be very similar and such animals may also therefore have a limited area range. The track made by some animals does not cover a large area but is more in a straight line and thus shows a linear range. If the annual range extends to the same point, animals exhibit a migration that might be to-and-fro or form a migration circuit.

A number of fish perform return migrations but like Pacific salmon, freshwater eels and lampreys, the circuit made by the individuals takes longer than one year. Some eels for example may take up to 20 years to complete their return migration. Baker (1982) terms these migration cycles as ontogenetic.

As fish move from place to place within their daily, monthly, annual or lifetime range, the question arises as to whether or not the animals know where they are going. If they do have a sense of location how and when did the various locations within the range become known? Even if some fish have a sense of location the problem arises as to how they have developed the mechanisms to find the different locations. The strategies and mechanisms of migrating from one range to another often become difficult to elucidiate. It is sometimes not difficult to appreciate why or how a species might be adaptable to certain environments but it is surprising when different individuals of that one species display vastly different migration patterns.

Sense of Location

Throughout the life of an animal various places will be visited and add to the lifetime range. However, some places in this so-called familiar area are more important and may thus be visited more than other areas. Other areas may be ranked low in suitability because the cost is too great to reach these ones or the resources within the area are no longer required. Each area that is familiar to an animal has various degrees of suitability depending on the resource required at a specific time. Familiarity with an area provides an animal with the opportunity to 'evaluate' or 'calculate' the 'cost' in terms of time, energy and risks required to migrate to that habitat. Movements within a familiar area are termed calculated migrations by Baker (1982). If an animal moves to a particular place within a familiar area, it does not have to waste or risk time and energy in non-calculated migrations.

Many fish appear to perform calculated migrations. Evidence for such movements comes from displacement experiments. If a fish is known to move between two locations regularly and continues to seek out one of these areas after being displaced, the fish has shown a preference and ability to locate such a habitat. Natural fish populations appear to stay within a familiar area. Thus, many trout spawn year after year in the same locations and return to the same feeding area until the next reproductive season. White bass, *Roccus chrysops,* return to specific lake shore areas to spawn. Many oceanodromous fish such as herring, cod, *Gadus morhua,* and plaice, *Pleuronectes platessa* perform ocean migration circuits to return to the same spawning grounds. It would thus appear that fish 'know' where they are and where they 'want' to go.

The central nervous system must certainly be involved in a sense of location. An adult salmon might thus recognize the odor of its natal stream and consequently follow that cue along an arduous migration route. The mechanism might involve a type of instinct and imprinting and not necessarily conscious cerebral awareness. It is difficult, however, to evaluate how a fish might obtain its sense of location. It is probably even more difficult to ascertain how a fish ascertains where or when to migrate.

Exploration

Once an area has become familiar, an animal can move from place to place very efficiently. The animal knows what part of the habitat has the resource it requires and how to get to that area with minimum cost and risk. The animal requires some form of spatial memory in order to accomplish movements within a familiar area and in addition must have achieved an adequate level of familiarization with its range.

Baker (1982) outlines the general sequence of events that must be undertaken by an animal in order to make some new habitat part of the familiar range. At birth the familiar range is zero and the animal must start building up new areas that it has become familiar with. This learning about new ideas includes movements into unfamiliar territory or so-called exploratory migrations. These migrations take the animal along unfamiliar routes to areas from which the animal will probably return to a familiar area. Once an animal has entered an unfamiliar area there will follow a period of habitat assessment. This involves an evaluation of suitability of the route or place itself with respect to the resources offered in comparison to other known areas. After undertaking an exploratory migration an animal must find its way back to a familiar area by some means of orientation or navigation. Migrations into new habitats are inefficient but necessary to add to a familiar area. Some excursions of this type may yield information about useful new habitats but may also lead to other areas that will be avoided, ignored or revisited for a future re-evaluation. If an animal moves to an unfamiliar area there must be some preferential decision to move out of a familiar area unless there has been some accidental displacement beyond the control of the animal.

Baker (1982) discusses a number of examples of both vertebrate and invertebrate animal species that seem to undertake exploratory migrations. A number of these animals make forays beyond their familiar areas and at some later stage of their life take up residence in the new area. Consequently some type of evaluation probably took place.

Exploratory migrations by some animals take place in the presence of adults. It thus appears that these animals learn from these migrations and can consequently build up their familiar areas from migrations with experienced adults. Some animals other than humans such as chimpanzees, other mammals and honey bees appear to be capable of social communication that is helpful in conveying information about exploratory migrations. Young fish, however, are likely only to learn by association with adults, but nevertheless such learning could be invaluable since it would make exploratory movements less inefficient and could remove many hazards.

Territory

Animals rarely move about in their habitat without encountering members of their own species. Individuals or groups are not always distributed at random in their environments but often occupy specific territories. The patterns of migration may be altered by these intraspecific interactions that take place due to the development of territories. Exploratory movements may be restricted or redirected due to territories. Also, the final area chosen and its size may also be influenced. The availability of resources and the number of individuals competing for those resources will affect territory size and freedom of movement in, out, through, or around certain areas.

As population density increases the dominant individuals will compete for the available resources in a particular area. If subordinate individuals experience aggressive behavior from a dominant individual, the choice of an area to settle is modified. To save energy it is advantageous for an explorer and territory resident to advertise their presence. When the density of a population increases the competition increases for the available resources and territories must be defended against intruders. The time spent in defense of a territory increases and a resident cannot go to overlapping areas; thus his territory will become smaller and movements will be more centrally located. In this type of situation, exploratory migrations out of necessity will be diminished.

Long-distance Migrations

An exploratory migration out of a familiar area might be easy to envisage if the distances traversed were relatively small. The extension of these exploratory movements over considerable distances is more difficult to appreciate, but the possibility exists that long-distance migrations of fish are no more than outward-bound journeys of the fish in search of a particular suitable habitat.

Many salmonid fish exhibit extended migrations into the ocean and return to their natal areas to spawn. Hasler (1983) has given a convincing review that the homing of some of these fish is due to an olfactory cue that the fish imprinted to at a young stage of its life cycle. However, not all spawning adults return to their natal areas to spawn. Between 10 to 20 per cent of most populations are 'lost' or 'stray' to alternative spawning locations. As Baker (1982) points out, these fish might not be lacking in orientation abilities but instead might be executing exploratory migrations in search of better spawning habitats.

Young salmon spend varying times in a stream environment and at that stage are territorial with some individuals being displaced from some habitats. Displacement studies show that some of the individuals display a home range fidelity. These young stream-dwelling salmonids exhibit movements upstream and downstream depending on the time of year. During these excursions it is possible that they are evaluating the different habitats for their suitability as future spawning sites. When young salmon are transferred to different streams at appropriate times in their early life, they will return as adults to the new environment and at an appropriate time of year for spawning in the new habitat to which they were displaced. This change in migration pattern and timing may have been learned or evaluated from the exposure to the new environment. Adults also appear to explore different streams before spawning and displaced individuals show a small percentage that spawn in sites other than their natal area. Supposedly unsuitable habitats due to a dam or the historical absence of spawning adults appear to be avoided by spawners. Restocking programmes of these areas have generally led to poor returns as well, which might be an indication that the fish when young assessed the habitat as potentially poor for future spawning. White (1936) describes the re-establishment of a salmon run in parts of the Apple River, Nova Scotia, after removal of a dam. The removal of the physical barrier did not encourage adults to return to the re-opened spawning area but as soon as young fry were introduced, spawning adults returned. The adults may have sensed the re-opened areas as potentially adequate for spawning because of the presence of young fish as indicators of a previously successful spawning.

Baker (1982) summarizes the movements of Atlantic herring from a number of different populations and during different stages in their life cycles. The immature herring appear to be much wider ranging which might be an indication of more exploratory movements before the older fish can establish a more rigorous and defined migration circuit. Other oceanic migrants such as the cod and the tunny, *Thunnus thynnus*, have expanded their migration circuits over the last few decades to include more northerly regions. Again, this may be an indication of exploratory behavior in which new habitats are evaluated. Many marine fish produce pelagic larvae that drift with ocean currents and indeed adults of some of these species also follow ocean currents to complete their migration circuits. Although passive drifting with a current is hard to visualize as being an exploratory movement, it does allow the fish (with the use of minimal energy) to enter and experience new habitats. If these young fish can store the information gained from these new areas, a repeatable migration pattern might be established.

Navigation

Fish may locate, assess and rank a new area by exploratory movements. Subsequently, this newly explored area might be added to the already present, familiar area. If the fish requires this new area later in its life cycle it must be capable of relocating it and recognizing the area once relocated. The new area might be recognized by a single or a combination of a number of characteristics. These characteristics might be due to visual cues, chemical cues, anomalies in the earth's magnetic field, temperature differences, salinity gradients, strengths of current directions, electric field changes, hydrostatic pressure differences, etc. Some of these cues or others may be assessed by particular species of fish and subsequently incorporated into some type of spatial memory, perhaps in association with some stable absolute axis such as compass bearings associated with celestial cues or the earth's magnetic field.

An animal might find its way back to a particular location within a familiar area by what Baker (1982) terms a route-based navigation. The animal may accomplish this by simply reversing the route of the outward journey. Another possibility is for the animal somehow to record all of the turns during an outward journey and calculate the correct bearing required to return directly. This latter type of route-based navigation (inertial navigation) would be more efficient if the outward journey was not linear. Another possibility for an animal to find its way back to a familiar area is to make use of characteristics of the new area instead of the route to the new area (location-based navigation). The new site would thus have features that had particular compass directions *relative* to the familiar area.

Location versus Direction

If an animal performs an exploratory migration and for one reason or another needs to return to a specific site in the familiar area, some type of direction sense is required for the animal to find its way back. Thus, although these animals have a sense of location and move from one location to another for various purposes, a sense of direction is also required. On the other hand, some animals may simply respond to a need to move in a certain direction irrespective of location. Daily or seasonal inshore offshore fish migrations or vertical migrations may be of this latter type. Animals that are short-lived and thus cannot afford to invest time in building up a familiar area may opt for this type of movement. Also, if suitable sites are numerous and individuals can move to them over short distance with relative ease, a sense of direction might also be more important than a sense of location.

Learning and Instinct

Animals appear to explore new areas and subsequently add these to a knowledge of a former familiar area. This behavior might even be a sequence of events leading to long-distance migrations. However, the question still arises as to whether this behavior is learned or whether it is part of a genetic make-up of the individual and is, therefore, only a display of innate or instinctual traits. As an animal moves through its environment it receives various sensory inputs from which it can assess the 'appropriateness' of that particular habitat. This process may be conscious or unconscious and may also rely on learning from others or from personal experience but may also be part of a 'pre-programming' or innate ability.

If an animal can build up a knowledge of a familiar area by a repeated series of explorations, some genetic predetermined 'desire' to explore must be present providing this behavior is innate. It does appear, from a number of experiments, that certain animals do exhibit periods of restlessness that might therefore be associated with exploratory behavior (see Baker, 1982). Furthermore, these periods of restlessness appear ontogenetically at the appropriate time as well as during the correct season. This observed restlessness in animals is not a random movement but in many cases is directed towards a specific compass direction. The displaying of an oriented restlessness can be modified in its timing by different environmental factors such as photoperiod and temperature. Migratory behavior not only requires appropriate orientation abilities to be developed but also many other processes. Thus, other mechanisms such as fattening, gonadal recrudescence and other anatomical modifications for migration also occur at desirable times and seem also to be controlled by endogenous changes that are of a pre-programmed nature.

Although it appears that many animals possess an innate ability to explore new areas and expand their familiar area and thus even accomplish long migrations, modifications of this behavior seem to be brought about by experience or learning from other individuals. Cross-breeding experiments with gulls and storks have indicated that the predetermined migratory patterns of certain individuals could be altered by learning from adults (see Baker, 1978). An animal appears to be given a programmed set of events through inheritance that occur in its morphology, physiology and behavior and which determine the characteristics of its lifetime track or migration. Depending on the species, this migration may be altered more or less by learning from individual experience, observation of other conspecifics or through communication.

All individuals within a given population do not inherit the same disposition for a particular form of migration. Some individuals may thus possess different innate tendencies to migrate in certain directions or migrate specific distances. Even an individual's genetic makeup may determine its ability to dominate in intraspecific competitions and thus determine when or if it will migrate. These individual differences in innate abilities or learned abilities, which in themselves may also be determined by innate characteristics, impart in a particular species a degree of variation with respect to migration. This variation in turn may lead to changes in migratory behavior or patterns in accordance with concomitant changes in that species' environment. Variations in different aspects of migration may be selected for or selected against and thus lead to evolutionary change in the same way as for any feature of an animal's make-up.

REFERENCES

Able, K.P. (1980) 'Mechanisms of orientation, navigation and homing' in S.A. Guthreaux, Jr, (ed.), *Animal Migration, Orientation, and Navigation*, Academic Press, New York.

Alderdice, D.F., J.R. Brett, D.R. Idler and U.H.M. Fagerlund (1954) 'Further observations on olfactory perception in migrating adult coho and spring salmon – Properties of the repellent in mammalian skin.' *Fish. Res. Board Can. Pac. Sta. Progr. Rep.*, *98*, 10-12.

Alverson, D.L. (1961) 'Ocean temperatures and their relation to albacore tuna (*Thunnus germo*) distribution in waters off the coast of Oregon, Washington and British Columbia.' *J. Fish. Res. Board Can.*, *18*, 145-52.

Arnold, G.P. (1974) 'Rheotropism in fishes.' *Biol. Rev.*, *49*, 515-76.

Aronson, L.R. (1951) 'Orientation and jumping behavior in the Gobiid fish *Bathygobius soparator*.' *Am. Mus. Novit.*, *1486*, 1-22.

Aronson, L.R. (1971) 'Further studies on orientation and jumping behavior in the gobiid fish, *Bathygobus soporator*.' *Ann. NY Acad. Sci.*, *188*, 378-407.

Ashby, K.R. (1957) 'The effect of steroid hormones on the brown trout (*Salmo trutta* L.) during the period of gonadal differentiation.' *J. Embryol. Exp. Morphol.*, *5*, 225-49.

Audet, C. and G. Chevalier (1981) 'Monoaminergic innervation of the caudal neurosecretory system of brook trout, *Salvelinus fontinalis* in relation to osmotic stimulation.' *Gen. Comp. Endocrinol.*, *45*, 189-203.

Averett, R.C. (1969) 'Influence of temperature on energy and material utilization by juvenile coho salmon.' PhD thesis, Oregon State University, Corvalis, Oregon.

Baggerman, B. (1975) 'An experimental study on the timing and breeding migration in the three-spined stickleback (*Gasterosteus aculeatus* L.).' *Archs. néerl., Zool.*, *12*, 1-213.

Baggerman, B. (1959) 'The role of external factors and hormones in migration of sticklebacks and juvenile salmon' in A. Gorbman (ed.), *Comparative Endocrinology*, Wiley, New York.

Baggerman, B. (1960) 'Salinity preference, thyroid activity and the seaward migration of four species of Pacific salmon.' *J. Fish. Res. Board Can.*, *17*, 295-322.

Baggerman, B. (1962) 'Some endocrine aspects of fish migration.' *Gen. Comp. Endocrinol. Suppl.*, *1*, 188-205.

Baggerman, B. (1963) 'The effect of TSH and antithyroid substances on salinity preference and thyroid activity in juvenile Pacific salmon.' *Can. J. Zool.*, *41*, 307-19.

Bailey, R.E. (1957) 'The effect of estradiol on serum calcium, phosphorus and protein of goldfish.' *J. Exp. Zool.*, *136*, 455-70.

Baker, R.R. (1969) 'The evolution of the migratory habit in butterflies.' *J. Anim. Ecol.*, *38*, 703-46.

Baker, R.R. (1978) *The Evolutionary Ecology of Animal Migration*, Holmes and Meier Publishers, Inc., New York/Hodder & Stoughton, London.

Baker, R.R. (1982) *Migration, Paths through Time and Space*, Hodder & Stoughton, London.

Balchen, J.G. (1976a) 'Principles of migration of fishes.' *SINTEF: The Engineering research foundation at the Technical University of Norway, Trondheim. Teknisk notat nr. 81 for NTNF/NFFR*, pp. 1-33.

Balchen, J.G. (1976b) 'Modelling of the biological state of fishes.' *SINTEF: Eng. Res. Found. Tech. Univ. Norway, Trondheim. Teknisk notat nr. 62 for NTNF/NFFR*, pp. 1-25.

Ball, J.N. (1965) 'Partial hypophysectomy in the teleost *Poecilia*: Separate identities of the teleostean growth hormone and teleostean prolactin-like hormone.' *Gen. Comp. Endocrinol.*, *5*, 654-61.

Ball, J.N., M. Olivereau, A.M. Slicher and K.D. Kallman (1965) 'Functional capacity of ectopic pituitary transplants in the teleost, *Poecilia formosa*, with a comparative discussion on the transplanted pituitary.' *Phil. Trans. Royal Soc. London, Ser. B. 249*, 69-99.

Ball, J.N., I. Chester-Jones, M.E. Forster, G. Hargreaves, E.F. Hawkins and K.P. Milne (1971) 'Measurement of plasma cortisol levels in the eel *Anguilla anguilla* in relation

to osmotic adjustments.' *J. Endocrinol, 50,* 75-96.
Bams, R.A. (1976) 'Survival and propensity for homing as affected by presence or absence of locally adapted paternal genes in two transplanted populations of pink salmon (*Oncorhynchus gorbuscha*).' *J. Fish. Res. Board Can., 33,* 2716-25.
Banks, J.W. (1969) 'A review of the literature on the upstream migration of adult salmonids.' *J. Fish. Biol., 1,* 85-136.
Baraduc, M.M. and M. Fontaine (1955) 'Etude comparée due métabolisme respiratoire de jeune saumon sédentaire (parr) et migrateur (smolt).' *C.R. Soc. Biol., 149,* 1327-9.
Barber, F.G. (1979) 'On ocean migration, speciation, cycle, dominance and density dependence in Pacific salmon.' *Fish. Mar. Ser. Tech. Report,* pp. 1-7.
Barbour, S.E. and E.T. Garside (1983) 'Some physiologic distinctions between freshwater and diadromous forms of the Atlantic salmon, *Salmo salar* L.' *Can. J. Zool., 61,* 1165-70.
Barlow, J.S. (1964) 'Inertial navigation as a basis for animal navigation.' *J. Theor. Biol., 6,* 67-117.
Barrington, E.J.W., N. Barron and D.J. Piggins (1961) 'The influence of thyroid power and thyroxine upon the growth of rainbow trout (*Salmo gairdneri*).' *Gen. Comp. Endocrinol., 1,* 170-8.
Bateson, P.P.G. (1979) 'How do sensitive periods arise and what are they for?' *Anim. Beh., 27,* 470-86.
Batschelet, E. (1965) *Statistical Methods for the Analysis of Problems in Animal Orientation and Certain Biological Rhythms*, Amer. Inst. Biol. Sci., Washington, DC.
Beamish, F.W.H. (1964) 'Influence of starvation on standard and routine oxygen consumption.' *Trans. Am. Fish. Soc., 93,* 103-7.
Beamish, F.W.H. (1978) 'Swimming capacity' in W.S. Hoar and D.J. Randall (eds.), *Fish Physiology, Vol. VII,* Academic Press, New York.
Beatty, D.D. (1966) 'A study of the succession of visual pigments in Pacific salmon (*Oncorhynchus*).' *Can. J. Zool., 44,* 429-56.
Beckett, J.S. (1973) 'Fish tagging.' *Underwater J.,* pp. 250-4.
Bennett, M.V.L. (1971) 'Electroreception' in W.S. Hoar and D.J. Randall (eds.), *Fish Physiology, Vol. V,* Academic Press, New York.
Bentley, P.J. (1971) 'Endocrines and osmoregulation. A comparative account of the regulation of water and salt in vertebrates' in D.S. Farner (ed.), *Zoophysiology and Ecology, vol. 1,* Springer-Verlag, New York.
Bentley, P.J. (1976) *Comparative Vertebrate Endocrinology*, Cambridge University Press, Cambridge.
Bentley, P.J. and B.K. Follet (1965) 'The effects of hormones on the carbohydrate metabolism of the lamprey *Lampetra fluviatilus*.' *J. Endocrinol., 31,* 127-37.
Berenbeim, D.Y., I. Durbrovin and E.M. Studenikina (1973) 'Forecast of the start of the fall migration of the anchovy, *Engraulis encrasicholus* (L.) through the Kerch Strait.' *J. Ichthyol., 13,* 313-17.
Berlind, A. (1973) 'Caudal neurosecretory system: A physiologists view.' *Amer. Zool., 13,* 759-70.
Bern, H.A. (1978) 'Endocrinological studies on normal salmon smoltification' in P.J. Gaillard and H.H. Boer (eds.), *Comparative Endocrinology*, Elsevier/North Holland Biomedical Press, Amsterdam.
Bernardi, C. (1948) 'Correlazioni dell'ipofisi e della tiroide con lo stato di maturazione della gonadi nelle *Anguille* gialle e Argentine.' *Riv. Biol., 40,* 186-228.
Bilinski, E. (1969) 'Lipid catabolism in fish muscle' in D.W. Neuhaus and J.E. Halver (eds.), *Fish in Research*, Academic Press, New York.
Billard, R., B. Breton, A. Fostier, B. Jalabert and C. Weil (1978) 'Endocrine control of the teleost reproductive cycle and its relation to external factors: salmonid and cyprinid models' in P.J. Gaillard and H.H. Boer (eds.), *Comparative Endocrinology*, Elsevier/North Holland Biomedical Press, Amsterdam.
Billard, R. and A.M. Escaffre (1973) 'Effects of HCG and carp gonadotropin on the maintainance of spermatogenesis in hypophysectomized goldfish (*Carassius auratus*).' *Int. Res. Commun. Syst., 73,* 12-15.
Billard, R., A. Fostier, C. Weil and B. Breton (1982) 'Endocrine control of spermatogenesis in teleost fish.' *Can. J. Fish. Aqua. Sci., 39,* 65-79.

Birnbaum, M.J., J. Schultz and J.N. Fain (1976) 'Hormone-stimulated glycogenolysis in isolated goldfish hepatocytes.' *Am. J. Physiol.*, *231*, 191-7.

Bitzukov, E.P. (1959) 'A contribution to the problem of vertical migrations of *Clupea harengus membras* L.' *Dokl. Akad. Nauk*. SSSR. Transl. *Biol. Sci.*, *128*, 766-9.

Black, E.C. and S.J. Tredwell (1967) 'Effect of a partial loss of scales and mucous on carbohydrate metabolism in rainbow trout (*Salmo gairdneri*).' *J. Fish. Res. Board Can.*, *24*, 939-53.

Blackburn, J.M. (1968) 'Digestive efficiency and growth in largemouth black bass.' MA thesis in Zoology, University of California, Davis, California.

Blem, C.R. (1980) 'The energetics of migration' in S.A. Gauthreaux Jr (ed.), *Animal Migration, Orientation, and Navigation*, Academic Press, New York.

Blüm, V. (1968) 'Immunological determination of injected mammalian prolactin in Cichlid fishes.' *Gen. Comp. Endocrinol.*, *11*, 595-602.

Blüm, V. and K. Fiedler (1965) 'Der Einflul von Prolactin auf das Brutplflegerverhalten von *Symphysodon aequifasciata axelrodi* Schultz.' *Die Naturwissenschaften*, *6*, 149-50.

Blüm, V. and K.M. Weber (1968) 'The influence of prolactin on the activity of steroid-3-β-ol dehydrogenase in the ovaries of the cichlid fish *Aequidens pulcher*.' *Experientia*, *24*, 1259.

Bock, F. (1928) 'Kastration und sekundäre Geschlechtsmerkmale bei Teleostiern.' *Z. wiss. Zool.*, *130*, 455-68.

Bodznick, D. (1975) 'The relationship of the olfactory EEG evoked by naturally occurring stream waters to the homing behavior of sockeye salmon (*Oncorhynchus nerka*, Walbaum).' *Comp. Biochem. Physiol.*, *52A*, 487-95.

Boëtius, I. and J. Boëtius (1967) 'Studies in the European eel *Anguilla anguilla* L. Experimental induction of the male sexual cycle, its relation to temperature and other factors.' *Meddr. Danm. Fisk.-og. Havunders.*, *Ny Ser.*, *4*, 339-405.

Boisseau, J.P. (1969) 'Prolactine et incubation chez l'Hippocampe.' *Colloq. Int. CNRS*, *177*, 205.

Bone, Q. (1966) 'On the function of the two types of myotomal muscle fibre in elasmobranch fish.' *J. Mar. Biol. Ass. UK*, *46*, 321-49.

Braemer, W. and H.O. Schwassmann (1963) 'Vom Rhythmus der Sonnenonentierung bei Fischen am Aquator.' *Ergeb. Biol.*, *26*, 181-201.

Braemer, W.A. (1960) 'A critical review of the sun-azimuth hypothesis.' *Cold Spring Harbour Symp. Quant. Biol.*, *25*, 413-28.

Brannon, E.L. (1967) 'Genetic control of migrating behavior of newly emerged sockeye salmon fry.' *Int. Pac. Salm. Fish. Comm. Bull. No. 16*.

Brannon, E.L. (1972) 'Mechanisms controlling migration of sockeye salmon fry.' *Int. Pac. Salm. Fish. Comm. Bull.*, *21*, 1-86.

Brannon, E.L. (1982) 'Orientation mechanisms of homing salmonids.' *Proc. Salmon Trout Migratory Behavior Symp.*, pp. 219-27.

Brannon, E.L., T.P. Quinn, G.L. Lucchetti and B.D. Ross (1981) 'Compass orientation of sockeye salmon fry from complex river system.' *Can. J. Zool.*, *59*, 1548-53.

Branover, G.G., A.S. Vasilyer, S.I. Gleizer and A.B. Tsinober (1971) 'A Study of the behavior of the eel in natural and aritificial and magnetic fields and an analysis of its receptor mechanism.' *J. Ichthyol.*, *11*, 604-8.

Breton, B., L. Horoszewicy, R. Billard and K. Bieniary (1980a) 'Temperature and reproduction in the tench: Effects of a rise in the annual temperature regime on gonadotropin level, gametogenesis and spawning. 1. The male.' *Reprod. Nutr. Dev.*, *20*, 105-18.

Breton, B., L. Horosyewicy, R. Billard and P. Epler (1980b) 'Temperature and reproduction in the tench: Effects of a rise in the annual temperature regime on gonadotropin level, gametogenesis and spawning. II. The female.' *Reprod. Nutr. Dev.*, *20*, 1011-24.

Breton, B., B. Jalabert and C. Weil (1975) 'Characterisation partielle d'un facteur hypothalamique de la liberation des hormones gonadotropes chez la Carpe (*Cyprinus Carpio*). Etude *in vitro*.' *Gen. Comp. Endocrinol.*, *25*, 405-15.

Brett, J.R. (1964) 'The respiratory metabolism and swimming performance of young sockeye salmon.' *J. Fish. Res. Board Can.*, *21*, 1183.

Brett, J.R. (1973) 'Energy expenditure of sockeye salmon, *Oncorhynchus nerka*, during sustained performance.' *J. Fish. Res. Board Can.*, *30*, 1799-809.

Brett, J.R. (1979) 'Environmental factors and growth' in W.S. Hoar, D.J. Randall and J.R. Brett (eds.), *Fish Physiology*, *Vol. VIII,* Academic Press, New York.

Brett, J.R. and N.R. Glass (1973) 'Metabolic rates and critical swimming speeds of sockeye salmon (*Oncorhynchus nerka*) in relation to size and temperature.' *J. Fish. Res. Board Can.*, *30*, 379-87.

Brett, J.R. and T.D.D. Groves (1979) 'Physiological Energetics' in W.S. Hoar, J.R. Randall and J.R. Brett (eds.), *Fish Physiology*, *Vol. VIII,* Academic Press, New York.

Brett, J.R. and D. MacKinnon (1954) 'Some aspects of olfactory perception in migrating adult coho and spring salmon.' *J. Fish. Res. Board Can.*, *11*, 310-18.

Brett, J.R. and D.B. Sutherland (1965) 'Respiratory metabolism of pumpkinseed (*Lepomis gibbosus*) in relation to swimming speed.' *J. Fish. Res. Board Can.*, *22*, 405-9.

Brett, J.R. and C.A. Zala (1975) 'Daily pattern of nitrogen excretion and oxygen consumption of sockeye salmon (*Oncorhynchus nerka*) under controlled conditions.' *J. Fish. Res. Board Can.*, *32*, 2479-86.

Brett, J.R., J.E. Shelbourn and C.T. Shoop (1969) 'Growth rate and body composition of fingerling sockeye salmon, *Oncorhynchus nerka*, in relation to temperature and ration size.' *J. Fish. Res. Board Can.*, *26*, 2363-94.

Brewer, K.J. and B.A. McKeown (1978) 'Effects of ergocryptine, prolactin, and growth hormone on survival and ion regulation in rainbow trout, *Salmo gairdneri*.' *Can. J. Zool.*, *56*, 2394-401.

Brewer, K.J. and B.A. McKeown (1980) 'Prolactin regulation in the coho salmon, *Oncorhynchus kisutch*.' *J. Comp. Physiol.*, *140*, 217-25.

Bridges, C.D. and C.E. Delisle (1974) 'Evolution of visual pigments.' *Exp. Eye. Res.*, *18*, 322-32.

Brocklesby, H.N. and O.F. Denstedt (1933) 'The industrial chemistry of fish oils with particular reference to those of British Columbia.' *Biol. Board Can. Bulletin*, *37*, 1-150.

Brocksen, R.W. and J.P. Burgess (1974) 'Preliminary investigations of the influence of temperature on food assimilation by rainbow trout, *Salmo gairdneri* Richardson.' *J. Fish. Biol.*, *6*, 93-7.

Brown, J.A. and J.A. Oliver (1984) 'The role of renin-angiotensin system in regulation of a single nephron function in trout' in B. Lofts and S.T.H. Chan (eds.), *Comparative Endocrinology*, Univ. Hong Kong Press, Hong Kong.

Buckland, F. (1880) *Natural History of British Fishes*, Unwin, London.

Bundie, H.F. (1977) 'Carbonic anhydrase.' *Comp. Biochem. Physiol.*, *57B*, 1-7.

Burzawa-Gerard, E. (1982) 'Chemical data on pituitary gonadotropins and their implication to evolution.' *Can. J. Fish. Aquat. Sci.*, *39*, 80-91.

Butler, D.G. (1968) 'Hormonal control of gluconeogenesis in the North American eel (*Anguilla rostrata*).' *Gen. Comp. Endocrinol, 10,* 85-91.

Butler, D.G. and F.J. Carmichael (1972) '(Na^+–K^+)-ATPase activity in eel (*Anguilla rostrata*) gills in relation to changes in environmental salinity. Role of adrenocortical steroids.' *Gen. Comp. Endocrinol, 19,* 241-7.

Butler, D.G., W.C. Clarke, E.M. Donaldson and R.W. Langford (1969) 'Surgical adrenalectomy of a teleost fish (*Anguilla rostrata* LeSeuer): effect on plasma cortisol and tissue electrolyte and carbohydrate and tissue electrolyte and carbohydrate concentrations.' *Gen. Comp. Endocrinol.*, *12*, 503-14.

Butler, M.J.A. (1982) 'Plight of the bluefin tuna.' *National Geographic, 162,* 220-39.

Byrne, J.E. (1968) 'The effects of photoperiod and temperature on the daily pattern of locomotor activity in juvenile sockeye salmon *Oncorhynchus nerka* (Walbaum).' PhD thesis, University of British Columbia, Vancouver, Canada.

Calaprice, J.R. (1971) 'X-ray spectrometric and multivariate analysis of sockeye salmon (*Oncorhynchus nerka*) from different geographical regions.' *J. Fish. Res. Board Can.*, *28*, 369-77.

Calaprice, J.R. and F.P. Calaprice (1970) 'Marking animals with micro-tags of chemical elements for identification by X-Ray spectroscopy.' *J. Fish. Res. Board Can.*, *27*, 317-30.

Calaprice, J.R., H.M. McSheffry and L.A. Lapi (1971) 'Radioisotope X-ray fluorescence spectrometry in aquatic biology: A review.' *J. Fish. Res. Board Can.*, *28*, 1583-94.

Campbell, C.M. (1978) '*In vitro* stimulation of vitellogenin incorporation into trout oocyte by salmon pituitary extracts.' *Ann. Biol. Anim. Biochem. Biophys.*, *18*, 1013-18.

Campbell, C.M. and D.R. Idler (1976) 'Hormonal control of vitellogenesis in hypophysectomized winter flounder (*Pseudopleuronectes americanus* Walbaum).' *Gen. Comp. Endocrinol.*, *28*, 143-50.
Campbell, C.M. and D.R. Idler (1977) 'Oocyte maturation and ovulation induced in hypophysectomized winter flounder (*Pseudopleuronectes americanus*) by preparation from pituitary glands of American plaice (*Hippoglossoides platessoides*).' *J. Fish. Res. Board Can.*, *34*, 2151-5.
Campbell, C.M., J.M. Walsh and D.R. Idler (1976) 'Steroids in the plasma of the winter flounder (*Pseudopleuronectes americanus* Walbaum). A seasonal study and investigation of steriod involvement in oocyte maturation.' *Gen. Comp. Endocrinol.*, *29*, 14-20.
Carlin, B. (1955) 'Tagging of salmon smolts in the River Lagan.' *Rep. Inst. Freshwater Res. Drottningholm*, *36*, 57-74.
Carlisle, D.B. and E.J. Denton (1959) 'On the metamorphosis of the visual pigments of *Anguilla anguilla* (L.).' *J. Mar. Biol. Ass. UK*, *38*, 97-102.
Carlson, C.A. and M.H. Shealy Jr (1972) 'Marking larval large-mouth bass with radiostrontium.' *J. Fish. Res. Board Can.*, *29*, 455-8.
Chan, D.K.O. (1972) 'Hormonal regulation of calcium balance in teleost fish.' *Gen. Comp. Endocrinol. Suppl.*, *3*, 411-20.
Chan, D.K.O. (1975) 'Cardiovascular and renal effects of urotensins I and II in the eel, *Anguilla rostrata*.' *Gen. Comp. Endocrinol.*, *27*, 52-61.
Chan, D.K.O. and H.A. Bern (1976) 'The caudal neurosecretory system: A critical evaluation of the two hormone hypothesis.' *Cell. Tiss. Res.*, *174*, 339-54.
Chan, D.K.O., I. Chester-Jones, I.W. Henderson and J.C. Rankin (1967) 'Studies on the experimental alteration of water and electrolyte composition of the eel (*Anguilla anguilla* L.).' *J. Endocrinol*, *37*, 297-317.
Chan, D.K.O., J.C. Rankin and I. Chester-Jones (1969) 'Influences of the adrenal cortex and the Corpuscles of Stannius on osmoregulation in the European eel (*Anguilla anguilla* L.) adapted to freshwater.' *Gen. Comp. Endocrinol. Suppl.*, *2*, 342-53.
Chang, J.P. and R.E. Peter (1983) 'Effects of dopamine on gonadotropin release in female goldfish, *Carassius auratus*.' *Neuroendocrinology*, *36*, 351-7.
Chang, V.M. and D.R. Idler (1960) 'Biochemical studies on sockeye salmon during migration. XII. Liver glycogen.' *Can. J. Biochem. Physiol.*, *38*, 553-8.
Chapman, D.W. (1962) 'Aggressive behavior in juvenile coho salmon as a cause of emigration.' *J. Fish. Res. Board Can.*, *19*, 1047-80.
Chester-Jones, I., I.W. Henderson, D.K.O. Chan, J.C. Rankin, W. Mosley, J.J. Brown, A.F. Lever, J.I.S. Robertson and M. Tree (1966) 'Pressor activity in extracts of the corpuscles of Stannius from the European eel (*Anguilla anguilla* L.).' *J. Endocrinol*, *34*, 393-408.
Chester-Jones, I., I.W. Henderson and W. Mosley (1964) 'Methods for the adrenalectomy of the European eel (*Anguilla anguilla* L.).' *J. Endocrinol.*, *30*, 155-6.
Chevalier, G. (1976) 'Ultrastructural changes in the caudal neurosecretory cells of the trout, *Salvelinus fontinalus* in relation to external salinity.' *Gen. Comp. Endocrinol.*, *29*, 441-54.
Chidambaram, S. (1972) 'Hormonal regulation of pigmentation, glycerine and natremia in the black bullhead, *Ictalurus melas*.' PhD thesis, University of Wisconsin, Madison.
Chidambaram, S., R.K. Meyer and A.D. Hasler (1972) 'Effects of hypophysectomy, pituitary autographs, prolactin, temperature and salinity of the medium on survival and natremia in the bullhead, *Ictalurus melas*.' *Comp. Biochem. Physiol.*, *43A*, 443-58.
Chidambaram, S., R.K. Meyer and A.D. Hasler (1973) 'Effect of hypophysectomy, isletectomy and ACTH on glycemia and hematocrit in the bullhead, *Ictalurus melas*.' *J. Exp. Zool.*, *184*, 75-80.
Cho, C.Y., S.J. Slinger and H.S. Bayley (1982) 'Bioenergetics of salmonid fishes: energy intake, expenditure and productivity.' *Comp. Biochem. Physiol.*, *73B*, 25-41.
Clarke, W.C. (1973) 'Sodium-retaining bioassay of prolactin in the intact teleost *Tilapia mossambica* acclimated to sea water.' *Gen. Comp. Endocrinol.*, *21*, 498-512.
Clarke, W.C. (1976) 'Effect of prolactin and growth hormone on growth, lipid content and seawater adaptation in juvenile sockeye salmon.' *Reg. Conf. Comp. Endocrinol.*,

Oregon State Univ., Corvalis Abstr., no. 38.
Clarke, W.C. and H.A. Bern (1980) 'Comparative endocrinology of prolactin' in C.H. Li, *Hormonal Proteins and Peptides, Vol. VIII*, Academic Press, New York.
Clarke, W.C. and Y. Nagahama (1977) 'Effect of premature transfer to sea water on growth and morphology of the pituitary, thyroid, pancreas, and interrenal in juvenile coho salmon (*Oncorhynchus kisutch*).' *Can. J. Zool.*, *55*, 1620-30.
Clarke, W.C. and J.E. Shelbourn (1977) 'Effect of temperature, photoperiod, and salinity on growth and smolting of underyearling coho salmon.' *Am. Zool.*, *17*, 957.
Clarke, W.C., S.W. Farmer and K.M. Hartwell (1977) 'Effect of teleost pituitary growth hormone on growth of *Tilapia mossambica* and on growth and seawater adaptation of sockeye salmon (*Oncorhynchus nerka*).' *Gen. Comp. Endocrinol.*, *33*, 174-8.
Clavert, J. and J.P. Zahnol (1956) 'Modifications hépatiques survenant pendant la vitellogenèse chez deux espèces de Poissons ovovivpares (*Xiphophorus helleri* et *Lebistes reticulatus*).' *C. R. Seanc. Soc. Biol.*, *150*, 1261-3.
Clemens, H.B. (1963) 'A model of albacore migration in the North Pacific Ocean.' *FAO Fish. Rep. No.* 6, *vol. 3*, *Experience Paper*, *31*, 1537-48.
Clemens, H.P., C. McDermitt and T. Inslee (1966) 'The effects of feeding methyltestosterone to guppies for 60 days after birth.' *Copeia No. 2*, 280-4.
Cohen, H. (1945) 'Effects of sex hormones on the development of the platyfish, *Platypoecilus maculatus*.' *Zoologica* (NY), *31*, 121-7.
Collins, G.B. (1952) 'Factors influencing the orientation of migrating anadromous fishes.' *US Fish. Wildl. Serv.*, *Fish. Bull.*, *52*, 375-96.
Conte, F.P. and H.H. Wagner (1965) 'Development of osmotic and ionic regulation in juvenile steelhead trout *Salmo gairdneri*.' *Comp. Biochem. Physiol.*, *14*, 603-20.
Cook, H. and A.P. van Overbeeke (1969) 'Ultrastructure of the eta cells in the pituitary gland of adult migratory sockeye salmon (*Oncorhynchus nerka*).' *Can. J. Zool.*, *47*, 937-41.
Cooper, J.C. and A.D. Hasler (1973) 'An electrophysiological approach to salmon homing – II.' *J. Fish. Res. Board Can.*, *Technical Report*, *415*, 1-44.
Cooper, J.C. and A.D. Hasler (1974) 'Electroencephalographic evidence for retention of olfactory cues in homing coho salmon.' *Science*, *183*, 336-8.
Cooper, J.C. and A.D. Hasler (1976) 'Electrophysiological studies of morpholine imprinted coho salmon (*Oncorhynchus kisutch*) and rainbow trout (*Salmo gairdneri*).' *J. Fish. Res. Board Can.*, *33*, 688-94.
Cooper, J.C., A.T. Scholz, R.M. Horrall, A.D. Hasler and D.M. Madison (1976) 'Experimental confirmation of the olfactory hypothesis with homing, artificially imprinted coho salmon (*Oncorhynchus kisutch*).' *J. Fish. Res. Board Can.*, *33*, 703-10.
Copp, D.H., E.C. Cameron, B. Cheney, A.G.F. Davidson and K.G. Henze (1962) 'Evidence for calcitonin – a new hormone from the parathyroid that lowers blood calcium.' *Endocrinology*, *70*, 638-49.
Cowey, C.B. and J.R. Sargent (1979) 'Nutrition' in W.S. Hoar, D.J. Randall and J.R. Brett (eds.), *Fish Physiology*, *Vol. VIII*, Academic Press, New York.
Cowey, C.B., J.A. Pope, J.W. Adron and A. Blair (1973) 'Studies on the nutrition of marine flatfish. The effect of oral administration of diethystilboestrol and cyproheptadine on the growth of *Pleuronectes platessa*.' *Mar. Biol.*, *19*, 1-6.
Craigie, E.H. (1926) 'A preliminary experiment on the relation of the olfactory sense to the migration of the sockeye salmon (*Oncorhynchus nerka* Walbaum).' *Trans. Roy. Soc. Can.*, *20*, 215-24.
Creach, Y. and A. Serfaty (1974) 'Le jeune et la realimentation chez la carpe (*Cyprinus carpio* L.).' *J. Physiol.*, *68*, 245-60.
Creutzberg, F. (1961) 'On the orientation of migrating elvers (*Anguilla vulgaris* Turt.) in a tidal area.' *Neth. J. Sea Res.*, *1*, 257-338.
Crim, L.W. (1982) 'Environmental modulation of annual and daily rhythms associated with reproduction in telost fishes.' *Can. J. Fish. Aqua. Sci.*, *39*, 17-21.
Crim, L.W. and D.M. Evans (1978) 'Seasonal levels of pituitary and plasma gonadotropin in male and female Atlantic salmon parr.' *Can. J. Zool.*, *56*, 1550-5.
Crim, L.W. and D.R. Idler (1978) 'Plasma gonadotropin, estradiol, and vitellogenin and gonad phosvitin levels in relation to the seasonal reproductive cycles of female brown

trout.' *Ann. Biol. Anim. Biochem. Biophys.*, *18*, 1001-5.
Dahlberg, M.L., D.L. Shumway and P. Doudoroff (1968) 'Influence of dissolved oxygen and carbon dioxide on swimming performance of largemouth bass and coho salmon.' *J. Fish. Res. Board Can.*, *25*, 49-70.
Dahlgren, V. (1914) 'On the electric motor nerve centers in the skates (Rajidae).' *Science*, *40*, 862-3.
Davitz, M.A. and K.R. McKaye (1978) 'Discrimination between horizontally and vertically polarized light by the cichlid fish *Pseudotropheus macrophythalmus*.' *Copeia*, *1978*, 333-4.
Delahunty, G., C.B. Schreck and V.L. de Vlaming (1977) 'Effect of pinealectomy, reproductive state, feeding regime and photoperiod on plasma cortisol in goldfish.' *Am. Zool.*, *17*, 873.
Demael-Suard, A., D. Garin, G. Brichon, M. Mure and G. Peres (1974) 'Neoglycogenese à partir de la glycine ^{14}C chez la tanche (*Tinca vulgaris* L.) au cours de l'asphyxie.' *Comp. Biochem. Physiol.*, *47A*, 1023-33.
Demski, L.S. and P.J. Hornby (1982) 'Hormonal control of fish reproductive behavior: brain-gonadel steroid interactions.' *Can. J. Fish. Aqua. Sci.*, *39*, 36-47.
de Ligny, W. and E.M. Pantelouris (1973) 'Origin of the European eel.' *Nature*, *246*, 518-19.
de Vlaming, V.L. (1972) 'The role of the endocrine system in temperature-controlled reproductive cycling in the estuarine fish, gobiid *Gillichthys mirabilis*.' *Comp. Biochem. Physiol.*, *41A*, 697-713.
de Vlaming, V.L. and M. Sage (1972) 'Diurnal variation in the fattening response to prolactin treatment in two cyprinodontial fishes (*Cyprinodon variegatus* and *Fundulus similis*).' *Contrib. Mar. Sci.*, *16*, 59-63.
de Vlaming, V.L. and B. I. Sundararaj (1972) 'Endocrine influences on the seminal vesicles in the estuarine gobiid fish, *Gillichthys mirabilis*.' *Biol. Bull.* (Woods Hole, Mass.), *142*, 243.
de Vlaming, V.L. and M.J. Vodicnik (1975) 'Effects of photoperiod-temperature regimes on pituitary gonadotrophs, pituitary gonadotropin potency and hypothalamic gonadotropin releasing activity in the teleost *Notemigonus crysoleucas*.' *J. Thermal. Biol.*, *1*, 119-25.
de Vlaming, V.L. and M.J. Vodicnik (1977) 'Effects of pinealectomy on pituitary gonadotrophs, pituitary gonadotropin potency and hypothalamic gonadotropin releasing activity in *Notemigonus crysoleucas*.' *J. Fish. Biol.*, 1073-86.
de Vlaming, V.L., M. Sage and B. Beitz (1975) 'Pituitary, adrenal and thyroid influences on osmoregulation in the euryhaline elasmobranch, *Dasyatis sabina*.' *Comp. Biochem. Physiol.*, *52A*, 505.
Dharmamba, M. and J. Maetz (1972) 'Effects of hypophysectomy and prolactin on the sodium balance of *Tilapia mossambica* in freshwater.' *Gen. Comp. Endocrinol.*, *19*, 175-83.
Dijkgraaf, S. and A.J. Kalmijn (1962) 'Verhaltungsversuche zur Funktion der Lorenzinischen Ampullen.' *Naturwissenschaften*, *49*, 400.
Dill, L.M., R.C. Ydenberg and A.H.G. Fraser (1981) 'Food abundance and territory size in juvenile coho salmon (*Oncorhynchus kisutch*).' *Can. J. Zool.*, *59*, 1801-9.
Dill, P.A. (1971) 'Perception of polarized light by yearling sockeye salmon (*Oncorhyncus nerka*).' *J. Fish. Res. Board Can.*, *28*, 1319-32.
Dingle, H. (1980) 'Ecology and evolution of migration' in S.A. Gauthreaux, Jr (ed.), *Animal Migration, Orientation, and Navigation*, Academic Press, New York.
Dizon, A.E., R.M. Horrall and A.D. Hasler (1973a) 'Long-term olfactory "memory" in coho salmon, *Oncorhynchus kisutch*.' *Fish. Bull. Calif.*, *71*, 315-17.
Dizon, A.E., R.M. Horrall and A.D. Hasler (1973b) 'Olfactory electroencephalographic responses of homing coho salmon, *Oncorhynchus kistuch*, to water conditioned by conspecifics.' *US Fish. Wildl. Serv. Fish. Bull.*, *71*, 893-6.
Dodd, J.M., P.A.C. Stuart-Kregar, J.P. Sumpter, L.W. Crim and R.E. Peter (1978) 'Premature sexual maturation in the Atlantic salmon (*Salmo salar* L.)' in P.J. Gaillard and H.H. Boer (eds.), *Comparative Endocrinology*, Elsevier/North Holland Biomedical Press, Amsterdam.

Dodson, J.J. and W.C. Leggett (1973) 'Behavior of adult American shad (*Alosa sapidissima*) homing to the Connecticut River from Long Island Sound.' *J. Fish. Res. Board Can.*, *30*, 1847-60.

Dodson, J.J. and W.C. Leggett (1974) 'Role of olfaction and vision in the behavior of American shad (*Alosa sapidissima*) homing to the Connecticut River from Long Island Sound.' *J. Fish. Res. Board Can.*, *31*, 1607-19.

Dodson, J.J. and J.C. Young (1977) 'Temperature and photoperiod regulation of rheotropic behavior in prespawning common shiners, *Notropis cornutus*.' *J. Fish. Res. Board Can.*, *34*, 341-6.

Donaldson, E.M. (1973) 'Reproductive endocrinology of fishes.' *Amer. Zool.*, *13*, 909-27.

Donaldson, E.M. and V.H.M. Fagerlund (1970) 'Effect of sexual maturation and gonadectomy at sexual maturity on cortisol secretion rate in sockeye salmon (*Oncorhynchus nerka*).' *J. Fish. Res. Board Can.*, *27*, 2287-96.

Donaldson, E.M., V.H.M. Fagerlund, D.A. Higgs and J.R. McBride (1979) 'Hormonal enhancement of growth' in W.S. Hoar, D.J. Randall and J.R. Brett (eds.), *Fish Physiology*, *Vol. VIII*, Academic Press, New York.

Donaldson, E.M., F. Yamazaki and W.C. Clarke (1968) 'Effect of hypophysectomy on plasma osmolarity in goldfish and its reversal by ovine prolactin and a preparation of salmon pituitary "prolactin".' *J. Fish. Res. Board Can.*, *25*, 1497-500.

Doneen, B.A. (1976) 'Water and ion movements in the urinary bladder of the gobiid teleost *Gillichthys marabilis* in response to prolactins and to cortisol.' *Gen. Comp. Endocrinol.*, *28*, 33-41.

Doneen, B.A. (1981) 'Effects of adaptation to sea water and to fresh water on activities and subcellular distribution of branchial Na^+–K^+–ATPase, low and high affinity Ca^{+}–$^{+}$ATPase, and ouabain – insensitive ATPase in *Gillichthys mirabilis*.' *J. Comp. Physiol.*, *145*, 51-61.

Døving, K.B., P. Enger and H. Nordeng (1973) 'Electrophysiological studies on the olfactory sense in char (*Salmo alpinus* L.).' *Comp. Biochem. Physiol.*, *45A*, 21-4.

Døving, K.B., H. Nordeng and B. Oakley (1974) 'Single unit discrimination of fish odours released by char (*Salmo alpinus* L.) populations.' *Comp. Biochem. Physiol.*, *47A*, 1051-63.

Driedzic, W.R. (1975) 'Energy metabolism in carp white muscle.' PhD thesis, University of British Columbia, Vancouver, Canada.

Driedzic, W.R. and P.W. Hochachka (1978) 'Metabolism in fish during exercise' in W.S. Hoar and D.J. Randall (eds.), *Fish Physiology*, *Vol. VIII*, Academic Press, New York.

Dubois, M.P., R. Billard, B. Breton and R.E. Peters (1979) 'Comparative distribution of somatostatin, LH-RH, neurophysin, and α-endorphin in the rainbow trout: an immunocytological study.' *Gen. Comp. Endocrinol.*, *37*, 220-32.

Duncan, D.W. and H.L.A. Tarr (1958) 'Biochemical studies on sockeye salmon during spawning migration. III. Changes in the protein and non-protein nitrogen fractions in muscles of migrating sockeye salmon.' *Can. J. Biochem. Physiol.*, *36*, 799-803.

Dupe, M. and Godet, R. (1969) 'Conditioning of the responsiveness of the central nervous system in the life cycle of the lungfish (*Protopterus annectens*).' *Gen. Comp. Endocrinol., Suppl. 2,* 278-83.

Durkin, J.T., W.J. Ebel and J.R. Smith (1969) 'A device to detect magnetized wire tags in migrating adult coho salmon.' *J. Fish. Res. Board Can.*, *26*, 3083-8.

Eales, J.G. (1965) 'Factors influencing seasonal changes in thyroid activity in juvenile steelhead trout, *Salmo gairdneri*.' *Can. J. Zool.*, *43*, 719-29.

Edwards, R.C.C., D.M. Finlayson and J.H.Steele (1972) 'An experimental study of the oxygen, growth, and metabolism of the cod (*Gadus morhua* L.).' *J. Exp. Mar. Biol. Ecol.*, *8*, 299-309.

Ege, R. and A. Krogh (1914) 'On the relation between the temperature and the respiratory exchange in fishes.' *Int. Gesamten Hydrobiol. Hydrogr.*, *1*, 48-55.

Elliott, J.M. (1976) 'Energy losses in the waste products of brown trout (*Salmo trutta* L.).' *J. Anim. Ecol.*, *45*, 561-80.

Emanuel, M.E. and J.J. Dodson (1979) 'Modification of the rheotropic behavior of male rainbow trout (*Salmo gairdneri*) by ovarian fluid.' *J. Fish. Res. Board Can.*, *36*, 63-8.

Enami, M. (1955) 'Caudal neurosecretory system in the eel (*Anguilla japonica*).' *Gunma J. Sci.*, *4*, 23-36.
Enomoto, Y. (1964) 'A preliminary experiment on the growth-promoting effect of growth hormone with thyroid-stimulating hormone and prolactin to the young rainbow trout (*Salmo irideus*).' *Nippon Suisan Gakkaishi*, *30*, 537-41.
Ensor, D.M. and J.N. Ball (1968) 'A bioassay for fish prolactin (paralactin).' *Gen. Comp. Endocrinol.*, *11*, 104.
Enright, J.T. (1972) 'When the beachhopper looks at the moon: the moon-compass hypothesis' in S.R. Galler, K. Schmidt-Koenig and R.E. Belleville (eds.), *Animal Orientation and Navigation*, *Scientific and Technical Information Office*, NASA, Washington, pp. 523-55.
Epstein, F.H., M. Cynamon and W. McKay (1971) 'Endocrine control of Na^{+}–K^{+}–ATPhase and seawater adaptation in *Anguilla rostrata*.' *Gen. Comp. Endocrinol.*, *16*, 323-8.
Evans, D.H., J.B. Claiborne, L. Farmer, C. Mallery and E.J. Krasny, Jr (1982) 'Fish gill ionic transport: methods and models.' *Biol. Bull.*, *163*, 108-30.
Eversole, W.J. (1941) 'The effects of pregeninolone and related steroids on sexual development in fish (*Lebistes reticulatus*).' *Endocrinology*, *28*, 603-10.
Evropeytseva, N.V. (1959) 'Transformation to smolt stage and downstream migration of young salmon.' *Fish. Res. Board Can.*, *Trans. Ser. No. 234*, 1-36.
Fagerlund, V.H.M. (1967) 'Plasma cortisol concentration in relation to stress in adult sockeye salmon during the freshwater stage of their life cycle.' *Gen. Comp. Endocrinol.*, *8*, 197-207.
Fagerlund, V.H.M. (1970) 'Response to mammalian ACTH of the interrenal tissue of sockeye salmon (*Oncorhynchus nerka*) at various stages of sexual maturation.' *J. Fish. Res. Board Can.*, *27*, 1169-72.
Farkas, T. (1967) 'Examinations of the fat metabolism in freshwater fishes – the sympathetic nervous system and the mobilization of fatty acids.' *Ann. Biol.*, *34*, 129-38.
Farkas, T. (1969) 'On control of fat mobilization in freshwater fishes.' *Acta Biochim. Biophys. Acad. Sci. Hung.*, *4*, 237-49.
Farmer, G.J. and F.W.H. Beamish (1969) 'Oxygen consumption of *Tilapia nilotica* in relation to swimming speed and salinity.' *J. Fish. Res. Board Can.*, *26*, 2807-21.
Farmer, S.W., H. Papkoff, T. Hayashida, T.A. Bewley, H.A. Bern and C.I. Li (1976) 'Purification of properties of teleost growth hormone.' *Gen. Comp. Endocrinol.*, *30*, 91-100.
Farmer, S.W., T. Hayashida, H. Papkoff and A.L. Polonev (1980) 'Characteristics of growth hormone isolated from sturgeon (*Acipenser güldenstädti*) pituitaries.' *Endocrinology*, *108*, 377-81.
Favre, L.C. (1960) 'Recherches sur l'adaption de Cyprins dorés à des milieux de salinité diverse et sur son déterminisme endocrinien.' Diplôme d'Etudes Supérieures, Paris.
Fenwick, J.C. (1974) 'The corpuscles of Stannius and calcium regulation in the North American eel (*Anguilla rostrata* Le Sueur).' *Gen. Comp. Endocrinol.*, *23*, 127-35.
Fenwick, J.C. (1976) 'Effect of staniectomy on calcium activated adenosinetriphosphatase activity in the gills of freshwater adapted North American eels, *Anguilla rostrata* Le Sueur.' *Gen. Comp. Endocrinol.*, *29*, 383-7.
Fenwick, J.C. (1979) Ca^{++}–activated adenosinetriphosphatase activity in the gills of freshwater- and seawater-adapted eels (*Anguilla rostrata*).' *Comp. Biochem. Physiol.*, *62b*, 67-70.
Fenwick, J.C. and Y.P. So. (1974) 'A perfusion study of the effect of stanniectomy on the net influx of calcium 45 across an isolated eel gill.' *J. exp. Zool.*, *188*, 125-31.
Fleming, W.R. and J.N. Ball (1972) 'The effect of prolactin and ACTH on the sodium metabolism of *Fundulus kansae* held in deionized water, sodium-enriched freshwater and concentrated seawater.' *Z. Vgl. Physiol.*, *76*, 125-34.
Folmar, L.C. and W.W. Dickhoff (1979) 'Plasma thyroxine and gill Na^{+}–K^{+}–ATPase changes during seawater acclimation of coho salmon, *Oncorhynchus kisutch*.' *Comp. Biochem. Physiol.*, *63*, 329-32.
Folmar, L.C. and W.W. Dickhoff (1980) 'The parr-smolt transformation (smoltification) and seawater adaptation.' *Aquaculture*, *21*, 1-37.
Folmar, L.C. and W.W. Dickhoff (1981) 'Evaluation of some physiological parameters as

predictive indices of smoltification.' *Aquaculture*, *23*, 309-24.

Fontaine, M. (1964) 'Corpuscles de Stannius et régulation ionique (Ca, K, Na) du milieu intérieur de l'anquille (*Anguilla anguilla* L.).' *C.R. Acad. Sci.*, *265*, 736-7.

Fontaine, M. (1975) 'Physiological mechanisms in the migration of marine and amphihaline fish.' *Adv. Mar. Biol.*, *13*, 241-355.

Fontaine, M. and J. Hatey (1950) 'Variations de la teneur du foie en glycogène chez le jeune saumon (*Salmo salar* L. au cours de la *smoltification*.' *C.R. Soc. Biol.*, *144*, 953-5.

Fontaine, M. and J. Hatey (1953) 'Contribution a l'Etude du Metabolisme glucidique du Saumon (*Salmo salar* L.) a diverses etapes de son development et de ses Migrations.' *Physiologia Comp. Oecol.*, *3*, 37-52.

Fontaine, M. and J. Hatey (1954) Sur la teneur en 17-hydroxycorticosteroides du plasma de Saumon (*Salmo salar* L.).' *C.R. Hebol. Seanc. Acad. Sci. Paris*, *239*, 319-21.

Fontaine, M., M. Baraduc and J. Hatey (1953) 'Influence de la thyroxinsation sur la teneur en glycogene du foie des poissons téléostéens.' *C.R. Soc. Biol.*, *147*, 214-16.

Forrest, J.N., A.D. Cohen, D.A. Schon and F.H. Epstein (1973) 'Na^+ transport and Na-K-ATPase in gills during adaptation to seawater: effects of cortisol.' *Amer. J. Physiol.*, *224*, 709-13.

Forskettl, J.K. and C. Scheffey (1982) 'The chloride cell: definitive identification as the salt secretory cell in teleosts.' *Science*, *215*, 164-6.

Forward, R.B., K.W. Horch and T.H. Waterman (1972) 'Visual orientation at the water surface by the teleost *Zenarchopterus*.' *Biol. Bull.*, *143*, 112-26.

Foster, R.C. (1975) 'Changes in urinary bladder and kidney function in the starry flounder (*Platichthys stellatus*) in response to prolactin and to freshwater transfer.' *Gen. Comp. Endocrinol.*, *27*, 153.

Fridberg, G. and H.A. Bern (1969) 'The urophysis and the caudal neurosecretory system of fishes.' *Biol. Rev.*, *43*, 175-99.

Fried, S.M., J.D. McCleave and G.W. LaBar (1978) 'Seaward migration of hatchery-reared Atlantic salmon, *Salmo salar*, smolts in the Penobscot River estuary, Maine: riverine movements.' *J. Fish. Res. Board Can.*, *35*, 76-87.

Fry, F.E.J. (1937) 'The summer migration of the cisco, *Leucichthys artedi* (LeSueur), in Lake Nipissing, Ontario.' *Univ. Tor. Stud. Biol. Ser. No. 44*, *Publ. Ont. Res. Lab.*, *No. 55*, 1-91.

Fry, F.E.J. (1957) 'The aquatic respiration of fish' in M.E. Brown (ed.), *The Physiology of Fishes*, *Vol. I*, Academic Press, New York.

Fry, F.E.J. (1971) 'The effect of environmental factors on the physiology of fish' in W.S. Hoar and D.J. Randall (eds.), *Fish Physiology*, *Vol. VI*, Academic Press, New York.

Fryer, J.N., N.Y.S. Woo, R.L. Gunther and H.A. Bern (1978) 'Effect of urophysial homogenates on plasma ion levels in *Gillichthys mirabilis* (Teleostei:Gobiidae).' *Gen. Comp. Endocrinol.*, *35*, 238-44.

Fujii, T. (1975) 'On the relation between the homing migration of the western Alaska sockeye salmon, *Oncorhynchus nerka* (Walbaum) and oceanic conditions in the Eastern Bering Sea.' *Mem. Fac. Fish. Hokkaido Univ.*, *22*, 99-192.

Gauthreaux, S.A., Jr (1980) *Animal Migration, Orientation and Navigation*, Academic Press, New York.

George, J.C. (1962) 'A histophysiological study of the red and white muscle of the mackerel.' *Am. Midl. Nat.*, *68*, 487-94.

Gibson, R.N. (1967) 'Studies on the movements of littoral fish.' *J. Anim. Ecol.*, *36*, 215-34.

Gill, T.S. and S.S. Khanna (1975) 'Effect of glucagon upon blood glucose level of the freshwater fish *Channa punctatus* (Block).' *Indian J. Exp. Biol.*, *13*, 298-300.

Gineste, P.J. (1955) 'Contribution à l'´etude de l'action des hormones hypophysaires sur la morphologie et le comportement de l'Anguille femelle.' *C.R. Seanc. Soc. Biol.*, *149*, 551-2.

Glova, G.J. and J.E. McInerney (1977) 'Critical swimming speeds of coho salmon (*Oncorhynchus kisutch*) fry to smolt stages in relation to salinity and temperature.' *J. Fish. Res. Board Can.*, *34*, 151-4.

Godwin, J.G., P.A. Dill and D.E. Drury (1974) 'Effects of thyroid hormones on behavior of yearling Atlantic salmon (*Salmo salar*).' *J. Fish. Res. Board Can.*, *31*, 1787-90.

Godsil, H.C. and E.K. Holmberg (1950) 'A comparison of the bluefin tunas, genus *Thunnus* from New England, Australia, and California.' *Calif. Div. Fish Game Fish. Bull.*, *77*, 55.

Goff, G.P. and J.M. Green (1978) 'Field studies of the sensory bias of homing and orientation to the home site in *Ulvaria subbifurcata* (Pisces:Stichaeidae).' *Can. J. Zool.*, *56*, 2220-4.

Goodyear, C.P. (1970) 'Terrestrial and aquatic orientation in the starhead topminnow, *Fundulus notti*.' *Science*, *168*, 603-5.

Goodyear, C.P. (1973) 'Learned orientation in the predator avoidance behavior of mosquitofish, *Gambusia affinis*.' *Behavior*, *45*, 191-224.

Goodyear, C.P. and D.H. Bennett (1979) 'Sun compass orientation of immature bluegill.' *Trans. Amer. Fish. Assoc.*, *108*, 555-9.

Goodyear, C.P. and D.E. Ferguson (1969) 'Sun-compass orientation in the mosquitofish, *Gambusia affinis*.' *Anim. Behav.*, *17*, 636-40.

Gorbman, A. (1969) 'Thyroid function and its control in fishes' in W.S. Hoar and D.J. Randall (eds.), *Fish Physiology*, *Vol. II*, Academic Press, New York.

Goswami, S.V., B. I. Sundararaj and E.M. Donaldson (1974) '*In vitro* maturation response of oocytes of the catfish *Heteropneustes fossilis* (Block) to salmon gonadotropin in ovary-head kidney coculture.' *Can. J. Zool.*, *52*, 745-8.

Gould, J.L., J.L. Kirschvink and K.S. Deffeyes (1978) 'Bees have magnetic remanence.' *Science*, *201*, 1026-8.

Grau, E.G. (1982) 'Is the lunar cycle a factor timing the onset of salmon migration?' *Proc. Salmon Trout Migratory Behavior Sym.*, pp. 184-9.

Grau, E.G. and M.H. Stetson (1977) 'Thyroidal responses to exogenous mammalian hormones in *Fundulus heteroclitus*.' *Am. Zool.*, *17*, 857.

Green, D.M. (1964) 'A comparison of stamina of brook trout from wild and domestic parents.' *Trans. Am. Fish. Soc.*, *93*, 96-100.

Greene, C.W. (1919) 'Biochemical changes in the muscle tissue of king salmon during the fast of spawning migration.' *J. Biol. Chem.*, *39*, 435-56.

Greene, C.W. (1926) 'The physiology of the spawning migration.' *Physiol. Rev.*, *6*, 201-41.

Greer Walker, M., F.R. Harden-Jones and G.P. Arnold (1978) 'Movements of plaice (*Pleuronectes platessa* L.) tracked in the open sea.' *J. Cons. Int. Explor. Mer.*, *38*, 72-100.

Greven, J.A.A., J.C.A. van der Meij and S.E. Wendelaar Bonga (1978) 'The relationship between hypocalcin and prolactin in the teleost (*Gasterosteus aculeatus*) in P. Gaillard and H. Boer (eds.), *Comparative Endocrinology*, Elsevier/North Holland Biomedical Press, Amsterdam.

Griffin, D.R. (1955) 'Bird navigation' in A. Watson (ed.), *Recent Studies in Avian Biology*, Univ. of Illinois Press, Urbana.

Groot, C. (1965) 'On the orientation of young sockeye salmon (*Oncorhynchus nerka*) during their seaward migration out of lakes.' *Behav. Suppl.*, *14*, 1-198.

Groot, C., K. Simpson, I. Todd, P.D. Murray and G.A. Buxton (1975) 'Movements of sockeye salmon (*Oncorhynchus nerka*) in the Skeena River estuary as revealed by ultrasonic tracking.' *J. Fish. Res. Board Can.*, *32*, 233-42.

Gross, W.L., P.O. Fromm and E.W. Roelofs (1963) 'Relationship between thyroid and growth in green sunfish, *Lepomis cyanellus* (Rafinesque).' *Trans. Am. Fish. Soc.*, *92*, 401-8.

Groves, A.B., G.B. Collins and P.S. Trefethen (1968) 'Roles of olfaction and vision in choice of spawning site by homing adult chinook salmon (*Oncorhynchus tshawytshcha*).' *J. Fish. Res. Board Can.*, *25*, 867-76.

Gunning, G.E. (1959) 'The sensory basis for homing in the longear sunfish, *Lepomis megalotis megalotis* (Rafinesque).' *Invest. Indiana Lakes Streams.*, *5*, 103-130.

Gunter, G. (1967) 'Some relationships of estuaries to the fisheries of the Gulf of Mexico' in G.H. Lauff (ed.), *Estuaries*, *Am. Assoc. Adv. Sci.*, Washington.

Hane, S. and O.H. Robertson (1959) 'Changes in plasma 17-hydroxycorticosteroids accompanying sexual maturation and spawning of the Pacific salmon (*Oncorhynchus tshawytscha* and rainbow trout (*Salmo gairdneri*).' *Proc. Natn. Acad. Sci. USA*, *45*, 886-93.

Hane, S., O.H. Robertson, B.C. Wexler and M.A. Krupp (1965) 'Adrenocortical response to stress and ACTH in Pacific salmon (*Oncorhynchus tshawytscha*) and steelhead trout (*Salmo gairdneri*) at successive stages in the sexual cycle.' *Endocrinology*, *78*, 791-800.

Hara, T.J. (1967) 'Electrophysiological studies of the olfactory system of the goldfish, *Carassius auratus* L. III. Effects of sex hormones on the electrical activity of the olfactory bulb.' *Comp. Biochem. Physiol.*, *22*, 209-26.
Hara, T.J. (1970) 'An electrophysiological basis for olfactory discrimination in homing salmon: A review.' *J. Fish. Res. Board Can.*, *27*, 565-86.
Hara, T.J., A. Gorbman and K. Veda (1966) 'Influence of thyroid state upon optically evoked potentials in midbrain of goldfish.' *Proc. Soc. Exp. Biol. Med.*, *122*, 471-5.
Hara, T.J., K. Ueda and A. Gorbman (1965) 'Electroencephalographic studies of homing salmon.' *Science*, *149*, 884-5.
Harden-Jones, F.R. (1957) 'Rotation experiments with blind goldfish.' *J. Exp. Biol.*, *34*, 259-75.
Harden-Jones, F.R. (1968) *Fish Migration*, Edward Arnold (Publishers) Ltd, London.
Hardisty, M.W. and I.C. Potter (1971) *The Biology of Lampreys, Vol. 1,* Academic Press, New York.
Hart, J.L. (1973) *Pacific Fishes of Canada*, Fisheries Research Board of Canada, Bulletin 180, Ottawa.
Hartman, G.F., B.C. Andersen and J.C. Scrivener (1982) 'Seaward movement of coho salmon (*Oncorhynchus kisutch*) fry in Carnation Creek, an unstable coastal stream in British Columbia.' *Can. J. Fish. Aquat. Sci.*, *39*, 588-97.
Hartman, W.L., W.R. Heard and D. Drucker (1967) 'Migratory behavior of sockeye salmon fry and smolts.' *J. Fish. Res. Board Can.*, *24*, 2069-99.
Hasler, A.D. (1954) 'Odour perception and orientation in fishes.' *J. Fish. Res. Board Can.*, *110*, 107-29.
Hasler, A.D. (1956) 'Perception of pathways by fish in migration.' *Quart. Rev. Biol.*, *31*, 200-9.
Hasler, A.D. (1960) 'Homing orientation in migrating fish.' *Ergebn. Biol.*, *23*, 94-115.
Hasler, A.D. (1966) *Underwater Guideposts*, University of Wisconsin Press, Madison.
Hasler, A.D. (1967) 'Animal orientation and navigation. Underwater guideposts for migrating fishes.' *Oregon State Univ. Biol. Colloq.*, *27*, 1-20.
Hasler, A.D. (1971) 'Orientation and fish migration' in W.S. Hoar and D.J. Randall (eds.), *Fish Physiology*, *Vol. VI,* Academic Press, New York.
Hasler, A.D. (1983) 'Synthetic chemicals and pheromones in homing salmon' in J.C. Rankin, T.J. Pitcher and R.T. Duggan (eds.), *Control Processes in Fish Physiology*, Croom Helm, London.
Hasler, A.D. and H.O. Schwassmann (1960) 'Sun orientation of fish at different latitudes.' *Cold Spring Harbour Symp. Quant. Biol.*, *25*, 429-41.
Hasler, A.D. and W.J. Wisby (1951) 'Discrimination of stream odors by fishes and relation to parent stream behavior.' *Am. Naturalist.*, *85*, 223-38.
Hasler, A.D., R.M. Horrall, W.J. Wisby and W. Braemer (1958) 'Sun-orientation and homing in fishes.' *Limnol. Oceanogr.*, *3*, 353-61.
Hasler, A.D., A.T. Scholz and R.M. Horrall (1978) 'Olfactory imprinting and homing in salmon.' *Am. Scient.*, *66*, 347-55.
Hatey, J. (1951) 'La fonction glycogenique du foie de l'anquille (*Anguilla anguilla* L.) aprés hypophysectomie.' *C.R. Seanc. Soc. Biol.*, *145*, 315-18.
Heape, W. (1931) *Emigration, Migration and Nomadism*, Heffer, Cambridge.
Henderson, I.W. and I. Chester-Jones (1967) 'Endocrine influences on the net extrarenal sodium fluxes of sodium and potassium in the European eel (*Anguilla anguilla* L.).' *J. Endocrinol, 37,* 319-25.
Henderson, I.W. and N.A.M. Wales (1974) 'Renal diuresis and anti-diuresis after injections of arginine vasotocin in the fresh-water eel (*Anguilla anguilla* L.).' *J. Endocrinol, 41,* 487-500.
Hickling, C.F. (1945) 'Marking fish with the electric tattooing needle.' *J. Mar. Biol. Ass. UK*, *26*, 166-9.
Higgs, D.A., E.M. Donaldson, H.M. Dye and J.R. McBride (1975) 'A preliminary investigation of the effect of bovine growth hormone on growth and muscle composition of coho salmon (*Oncorhynchus kisutch*).' *Gen. Comp. Endocrinol.*, *27*, 240-53.
Higgs, D.A., E.M. Donaldson, H.M. Dye and J.R. McBride (1976) 'Influence of bovine growth hormone and L-thyroxine on growth, muscle composition, and histological

structure of the gonads, thyroid, pancreas, and pituitary of coho salmon (*Oncorhynchus kisutch*).' *J. Fish. Res. Board Can.*, *33*, 1585-603.
Higgs, D.A., V.H.M. Fagerlund, J.R. McBride, H.M. Dye and E.M. Donaldson (1977) 'Influence of combinations of bovine growth hormone, 17α-methyltestosterone and L-thyroxine on growth of yearling coho salmon (*Oncorhynchus kistuch*).' *Can. J. Zool.*, *55*, 1048-56.
Higgs, D.A., E.M. Donaldson, J.R. McBride and H.M. Dye (1978) 'Evaluation of the potential for using a chinook salmon (*Oncorhynchus tshawytscha*) pituitary extract versus bovine growth hormone to enhance the growth of coho salmon (*Oncorhynchus kisutch*).' *Can. J. Zool.*, *56*, 1226-31.
Hirano, T. (1969) 'Effects of hypophysectomy and salinity change on plasma cortisol concentration in the Japanese eel, *Anguilla japonica*.' *Endocrinol Japon., 16*, 557-60.
Hirano, T. (1975) 'Effects of prolactin on osmotic and diffusion permeability of the urinary bladder of the flounder, *Platichthys flesus*.' *Gen. Comp. Endocrinol.*, *27*, 88-94.
Hirano, T., M. Morisawa, M. Ando and S. Utida (1975) 'Adaptive changes in ion and water transport mechanisms in the eel intestine' in J.W.L. Robinson (ed.), *Intestinal Ion Transport*, MTP Med. Tech. Publ., Lancaster.
Ho, S.M. and D.K.O. Chan (1980) 'Branchial ATPases and ionic transport in the eel *Anguilla japonica*. II. Ca^{++}–ATPase.' *Comp. Biochem. Physiol.*, *67B*, 639-45.
Ho, F.C.W. and W.E. Vanstone (1961) 'Effect of estradiol monobenzoate on some serum constituents of maturing sockeye salmon (*Oncorhynchus nerka*).' *J. Fish. Res. Board Can.*, *18*, 859-64.
Hoar, W.S. (1939) 'The thyroid gland of the Atlantic salmon.' *J. Morphol.*, *65*, 257-95.
Hoar, W.S. (1951) 'Hormones in fish.' *Publ. Ont. Fish. Res. Lab. Biol. Stud. No. 71*, 1-51.
Hoar, W.S. (1956) 'The bahavior of migrating pink and chum salmon fry.' *J. Fish. Res. Board Can.*, *13*, 309-25.
Hoar, W.S. (1958) 'Effects of synthetic thyroxine and gonadal steroids on the metabolism of goldfish.' *Can. J. Zool.*, *36*, 113-21.
Hoar, W.S. (1965) 'The endocrine system as a chemical link between the organism and its environment.' *Trans. R. Soc. Can. Ser.*, *4*, *3*, 175-200.
Hoar, W.S. (1969) 'Reproduction' in W.S. Hoar and D.J. Randall (eds.), *Fish Physiology*, *Vol. III*, Academic Press, New York.
Hoar, W.S. (1976) 'Smolt transformation: evolution, behavior and physiology.' *J. Fish. Res. Board Can.*, *33*, 1234-52.
Hoar, W.S., M.H. Keenleyside and R.G. Goodall (1955) 'The effects of thyroxine and gonadal steroids on the activity of salmon and goldfish.' *Can. J. Zool.*, *33*, 428-39.
Hoar, W.S., M.H.A. Keenleyside and R.G. Goodall (1957) 'Reactions of juvenile pacific salmon to light.' *J. Fish. Res. Board Can.*, *14*, 815-30.
Hoar, W.S., D. MacKinnon and A. Redlich (1952) 'Effects of some hormones on the behavior of salmon fry.' *Can. J. Zool.*, *30*, 73-286.
Hochachka, P.W. (1962) 'Thyroidal effects on pathways for carbohydrate metabolism in a teleost.' *Gen. Comp. Endocrinol.*, *2*, 499-505.
Holmes, W.N. and I.M. Stanier (1966) 'Studies on the renal exretion of electrolytes by the trout (*Salmo gairdneri*).' *J. Exp. Biol.*, *44*, 33-46.
Honma, Y. (1959) 'Studies on the endocrine glands of the salmonid fish, the Ayu, *Plecoglossus altivelis* Temminck et Schlegel. I. Seasonal variation in the endocrines of the annual fish.' *J. Fac. Sci., Niigata Univ., Ser. II.*, *2*, 223-33.
Honma, Y. and E. Tamura (1963) 'Studies on the endocrine glands of the salmonid fish, the Ayu, *Plecoglossus altivelis*. V. Seasonal changes in the endocrines of the land-locked form, the Koayu.' *Zoologica.*, *48*, 25-32.
Hontela, A. and R.E. Peter (1980) 'Effects of pinealectomy, blinding, and sexual condition on serum gonadotropin levels in the goldfish.' *Gen. Comp. Endocrinol.*, *40*, 168-79.
Horn, G. (1979) 'Imprinting – in search of neural mechanisms.' *Trends in Neurosciences*, *2*, 219-22.
Houston, A.H. (1957) 'Response of juvenile chum, pink and coho salmon to sharp sea water gradients.' *Can. J. Zool.*, *35*, 371-83.
Houston, A.H. (1960) 'Variations in the plasma level of chloride in hatchery-reared yearling Atlantic salmon during parr-smolt transformation and following transfer into sea

water.' *Nature*, *185*, 632-3.
Hudson, R.C.L. (1973) 'On the function of the white muscles in teleosts at intermediate swimming speeds.' *J. Exp. Biol.*, *58*, 509-22.
Hughes, R.B. (1963) 'Chemical studies on herring (*Clupea harengus*) 7. Collagen and cohesiveness in heat-processed herring and observations on a seasonal variation in collagen content.' *J. Sci. Fd. Agric.*, *14*, 432-41.
Hurley, D.A. and W.L. Woodall (1968) 'Responses of young pink salmon to vertical temperature and salinity gradients.' *Prog. Rep. Int. Pac. Salm. Fish. Comm.*, *19*, 1-80.
Hynes, H.B.N. (1970) *The Ecology of Running Waters*, Liverpool University Press, Liverpool.
Idler, D.R. and I. Bitners (1958) 'Biochemical studies on sockeye salmon during spawning migration. V. Cholesterol, fat, protein and water in the body of the standard fish.' *J. Fish. Res. Board Can.*, *16*, 235-41.
Idler, D.R. and W.A. Clemens (1959) 'The energy expenditure of Fraser River sockeye salmon during the spawning migration to Chilko and Stuart Lakes.' *Int. Pac. Salm. Fish. Comm.*, *New Westminster, BC, Rept. No. 6*, 1-80.
Idler, D.R. and H.C. Freeman (1967) 'Protein binding of steroids in fish plasma.' *Gen. Comp. Endocrinol.*, *9*, 459.
Idler, D.R. and T.B. Ng (1979) 'Studies on two types of gonadotropins from both salmon and carp pituitaries.' *Gen. Comp. Endocrinol.*, *38*, 421-40.
Idler, D.R., H.C. Freeman and B. Truscott (1964) 'Steroid hormones in the plasma of spawned Atlantic salmon, *Salmo salar*, and a comparison of their determination by biological and chemical assay methods.' *Can. J. Biochem. Physiol.*, *42*, 211-18.
Idler, D.R., D.A. Horne and G.B. Sangalang (1971) 'Identification and quantification of the major endrogons in testicular and peripheral plasma of Atlantic salmon (*Salmo salar*) during sexual maturation.' *Gen. Comp. Endocrinol.*, *16*, 257-67.
Idler, D.R., A.P. Ronald and P.J. Schmidt (1959) 'Biochemical studies on sockeye salmon during spawning migration. VII. Steroid hormones in plasma.' *Can. J. Biochem. Physiol.*, *37*, 1227-38.
Idler, D.R., K.M. Shamsuzzaman and M.P. Burton (1978) 'Isolation of prolactin from salmon pituitary.' *Gen. Comp. Endocrinol.*, *35*, 409-18.
Idler, D.R., V.H.M. Fagerlund and H. Mayok (1956) 'Olfactory perception in migrating salmon. I. L-serine, a salmon repellent in mammalian skin.' *J. Gen. Physiol.*, *39*, 889-92.
Idler, D.R., J.R. McBride, R.E.E. Jonas and N. Tomlinson (1961) 'Olfactory perception in migrating salmon. II. Studies on a laboratory bioassay for homestream water and mammalian repellent.' *Can. J. Biochem. Physiol.*, *39*, 1575-84.
Ince, B.W. and A. Thorpe (1974) 'Effects of insulin and of metabolite loading on blood metabolites in the European silver eel, *Anguilla anguilla* L.' *Gen. Comp. Endocrinol.*, *23*, 460-71.
Ince, B.W. and A. Thorpe (1976) 'The *in vivo* metabolism of ^{14}C-glucose and ^{14}C-glycine in insulin-treated northern pike (*Esox lucius* L.).' *Gen. Comp. Endocrinol.*, *28*, 481-6.
Inui, Y., S. Arai and M. Yokote (1975) 'Gluconeogenesis in the eel. VI. Effects of hepatectomy, alloxan, and mammalian insulin on the behavior of plasma amino acids.' *Nippon Suisan Gakkaishi, 41(11)*, 1105-11.
Jakim, E. and G. LaRoche (1973) 'Protein synthesis in *Fundulus heteroclitus* muscle.' *Comp. Biochem. Physiol.*, *A44*, 851-66.
Jalabert, B. (1976) '*In vitro* oocyte maturation and ovulation in rainbow trout (*Salmo gairdneri*), northern pike (*Esox lucius*) and goldfish (*Carassius auratus*).' *J. Fish Res. Board Can.*, *33*, 974-88.
Jefferts, K.B., P.K. Bergman, and H.F. Fiscus (1963) 'A coded wire identification system for macroorganisms.' *Nature*, *198*, 460-2.
Johnson, L. (1966) 'Experimental determination of food consumption of pike, *Esox lucius*, for growth and maintenance.' *J. Fish. Res. Board Can.*, *23*, 1495-505.
Johnston, C.E. and J.G. Eales (1967) 'Purines in the integument of the Atlantic salmon (*Salmo salar*) during parr-smolt transformation.' *J. Fish. Res. Board Can.*, *24*, 955-64.
Johnsen, P.B. and A.D. Hasler (1980) 'The use of chemical cues in the upstream migration of coho salmon, *Oncorhynchus kisutch* Walbaum.' *J. Fish. Biol.*, *17*, 67-73.
Johnston, I.A. and G. Goldspink (1973) 'Some effects of prolonged starvation on the

metabolism of the red and white myotomal muscles of the plaice (*Pleuronectes platessa*).' *Mar. Biol.*, *19*, 348-53.
Jonas, R.E.E. and R.A. MacLeod (1960) 'Biochemical studies on sockeye salmon during spawning migration. X. Glucose, total protein, non-protein nitrogen and amino acid nitrogen in plasma.' *J. Fish. Res. Board Can.*, *17*, 125-6.
Jones, D.R. (1971) 'Theoretical analysis of factors which may limit the maximum oxygen uptake of fish: the oxygen cost of the cardiac and branchial pumps.' *J. Theor. Biol.*, *32*, 341-9.
Jones, D.R. and D.J. Randall (1978) 'The respiratory and circulatory systems during exercise' in W.S. Hoar and D.J. Randall (eds.), *Fish Physiology, Vol. VII*, Academic Press, New York.
Joseph, M.M. and A.H. Meier (1971) 'Daily variations in the fattening response to prolactin in *Fundulus gradis* held on different photoperiods.' *J. Exp. Zool.*, *178*, 59-62.
Kalmijn, A.J. (1966) 'Electroperception in sharks and rays.' *Nature*, *212*, 1232-3.
Kalmijn, A.J. (1971) 'The electric sense of sharks and rays.' *J. Exp. Biol.*, *55*, 371-83.
Kalmijn, A.J. (1978) 'Experimental evidence of geomagnetic orientation in elasmobranch fisheries' in K. Schmidt-Koenig and W.T. Keeton (eds.), *Animal Migration, Navigation and Homing*, Springer-Verlag, Berlin.
Kalmus, H. (1964) 'Comparative physiology: Navigation by animals.' *Ann. Rev. Physiol.*, *26*, 109-30.
Kamra, S.K. (1966) 'Effect of starvation and refeeding on some liver and blood constituents of Atlantic cod (*Gadus morhua* L.).' *J. Fish. Res. Board Can.*, *23*, 975-82.
Kawauchi, H., K. Abe, A. Takahashi, T. Hirano, S. Hasegawa, N. Naito and Y. Nakai (1983) 'Isolation and properties of chum salmon prolactin.' *Gen. Comp. Endocrinol.*, *49*, 446-58.
Keeton, W.T. (1974) 'The orientational and navigational basis of homing in birds.' *Adv. Study Behav.*, *5*, 47-132.
Kelso, B.W., T.G. Northcote and C.F. Wehrhahn (1981) 'Genetic and environmental aspects of the response to water current by rainbow trout (*Salmo gairdneri*) originating from inlet and outlet streams of two lakes.' *Can. J. Zool.*, *59*, 2177-85.
Khoo, H.W. (1974) 'Sensory basis of homing in the intertidal fish *Oligocothus maculosus*, Girard.' *Can. J. Zool.*, *52*, 1023-9.
Kiceniuk, J.W. and D.R. Jones (1977) 'The oxygen transport system in trout (*Salmo gairdneri*) during sustained exercise.' *J. Exp. Biol.*, *69*, 247.
King, J.A. and R.P. Millar (1979) 'Heterogeneity of vertebrate luteinizing hormone-releasing hormone.' *Science*, *206*, 67-9.
King, J.A. and R.P. Millar (1980) 'Comparative aspects of luteinizing hormone-releasing hormone structure and function in vertebrate phylogeny.' *Endocrinology*, *106*, 707-17.
Klar, G.T., C.B. Stalnaker and T.M. Farley (1979) 'Comparative physical and physiological performance of rainbow trout, *Salmo gairdneri*, of distinct lactate dehydrogenase B^2 phenotype.' *Comp. Biochem. Physiol.*, *63A*, 229-35.
Kleerekoper, H. (1969) *Olfaction in Fishes*, Indiana University Press, Bloomington.
Kleerekoper, H., A.M. Timms, G.F. Westlake, F.B. Davy and T. Malar (1969) 'Inertial guidance system in the orientation of the goldfish (*Carassius auratus*).' *Nature*, *223*, 501-2.
Koch, H.J. (1968) 'Migration' in E.J.W. Barrington and C.B. Jorgenson (eds.), *Perspectives in Endocrinology*, Academic Press, New York.
Komourdjian, M.P., R.L. Saunders and J.C. Fenwick (1976a) 'Evidence for the role of growth hormone as a part of a "light-pituitary axis" in growth and smoltification of Atlantic salmon (*Salmo salar*).' *Can. J. Zool.*, *54*, 544-51.
Komourdjian, M.P., R.L. Saunders and J.C. Fenwick (1976b) 'The effect of procine somatotropin on growth and survival of Atlantic salmon (*Salmo salar*).' *Can. J. Zool.*, *54*, 534-5.
Kondo, H., Y. Hirano, N. Nakayama and M. Miyake (1963) 'Offshore distribution and migration of Pacific salmon (genus *Oncorhynchus*) based on tagging studies (1958-1961).' *Int. N. Pac. Fish. Comm. Bull.*, *17*, 1-128.
Kramer, K.L., M.P. Schreibman and P.K.T. Pang (1973) 'The effect of 2-Br-alpha-ergocryptine-methane sulfonate on fresh water survival and pituitary cytology in the tele-

ost, *Fundulus heteroclitus*.' *Am. Zool.*, *13*, 1279.

Kriebel, R.M. (1980) 'Ultrastructural changes in the urophysis of *Molliensia sphenops* following adaptation to sea water.' *Cell. Tiss. Res.*, *207*, 135-42.

Krogh, A. (1914) 'The quantitative relation between temperature and standard metabolism in animals.' *Int. Z. Phys.-Chem. Biol.*, *1*, 491-508.

Krueger, H.M., J.B. Saddler, G.A. Chapman, I.J. Tinsley and R.R. Lowry (1968) 'Bioenergetics, exercise and fatty acids of fish.' *Am. Zool.*, *8*, 119-29.

Kutty, M.N. (1968) 'Influence of ambient oxygen on the swimming performance of goldfish and rainbow trout.' *Can. J. Zool.*, *46*, 647-53.

Kutty, M.N. (1972) 'Respiratory quotient and ammonia excretion in *Tilapia mossambica*.' *Mar. Biol.*, *16*, 126-33.

Kutty, M.N. and R.L. Saunders (1973) 'Swimming performance of young Atlantic salmon (*Salmo salar*) as affected by reduced ambient oxygen concentration.' *J. Fish. Res. Board Can.*, *30*, 223-7.

Laevastu, T. and I. Hela (1970) *Fisheries Oceanography: New Ocean Environmental Services*, Fishing New Books Ltd, London.

Laevastu, T. and H. Rosa Jr (1963) 'Distribution and relative abundance of tuna in relation to their environment.' *FAO Fish. Rep.*, *3*, 1835-51.

Lahlou, B. and A. Giordan (1970) 'Le contrôle hormonal des echanges et de la balance de l'eau chez le Téléostéen d'eau douce *Carassius auratus*, intact et hypophysectomisé.' *Gen. Comp. Endocrinol.*, *14*, 491.

LaLanne, J.J. and G. Safsten (1969) 'Age determination from scales of chum salmon (*Oncorhynchus keta*).' *J. Fish. Res. Board Can.*, *26*, 671-81.

Lam, T.J. (1969) 'Effect of prolactin on loss of solutes via the head region of the early winter marine threespine stickleback (*Gasterosteus aculeatus*).' *Can. J. Zool.*, *47*, 865-70.

Lam, T.J. (1972) 'Prolactin and hydromineral regulation in fishes.' *Gen. Comp. Endocrinol. Suppl.*, *3*, 328-38.

Lam, T.J. and W.S. Hoar (1967) 'Seasonal effects of prolactin on freshwater osmoregulation of the marine form (*Trachurus*) of the stickleback, *Gaserosteus aculeatus*.' *Can. J. Zool.*, *45*, 509-16.

Lam, T.J. and J.F. Leatherland (1969) 'Effect of prolactin on freshwater survival of the marine form of the three-spine stickleback, *Gasterosteus aculeatus*, in the early winter.' *Gen. Comp. Endocrinol.*, *12*, 385-7.

Lam, T.J. and J.F. Leatherland (1970) 'Effect of hormones on survival of the marine form (*Trachurus*) of the three-spine stickleback (*Gasterosteus aculeatus* L.) in deionized water.' *Comp. Biochem. Physiol.*, *33*, 295-302.

Landgrebe, F.W. (1941) 'The role of the pituitary and the thyroid in the development of teleosts.' *J. Exp. Biol.*, *18*, 162-9.

Landsborough Thompson, A. (1942) *Bird Migration*, 2nd edn, Witherby, London.

LaRoche, G., A.N. Woodall, C.L. Johnson and J.E. Halver (1966) 'Thyroid function in the rainbow trout (*Salmo gairdneri* Rich.). II. Effects of thyroidectomy on the development of young fish.' *Gen. Comp. Endocrinol.*, *6*, 249-66.

Larsson, A.L. (1973) 'Metabolic effects of epinephrine and norepinephrine in the eel *Anguilla anguilla* L.' *Gen. Comp. Endocrinol.*, *20*, 155-67.

Leatherland, J.F. (1970) 'Seasonal variation in the structure and ultrastructure of the pituitary gland in the marine form (*Trachurus*) of the threespine stickleback, *Gasterosteus aculeatus* L. I. Rostral pars distalis.' *Z. Zellforsch.*, *104*, 301-17.

Leatherland, J.F. and M. Hyder (1975) 'Effect of thyroxine on the ultrastructure of the hypophyseal proximal pars distalis in *Tilapia zillii*.' *Can. J. Zool.*, *53*, 686-90.

Leatherland, J.F. and B.A. McKeown (1974) 'Effect of ambient salinity on prolactin and growth hormone secretion and on hydromineral regulation in kokanee salmon smolts (*Oncorhynchus nerka*).' *J. Comp. Physiol.*, *89*, 215-26.

Leatherland, J.F., B.A. McKeown and T.M. John (1974) 'Circadian rhythm of plasma prolactin, glucose and free fatty acid in juvenile kokanee salmon, *Oncorhynchus nerka*.' *Comp. Biochem. Physiol.*, *47*, 821-8.

Lederis, K. (1973) 'Current studies on urotensins.' *Amer. Zool.*, *13*, 771-3.

Lederis, K. (1977) 'Chemical properties and the physiological and pharmacological actions of urophysial peptides.' *Amer. Zool.*, *17*, 823-32.

Lederis, K., A. Letter and G. Moore (1976) 'Chemical and pharmacological properties of urotensins I and II.' *Gen. Comp. Endocrinol.*, *29*, 245-6.

Lee, R.W. and A.H. Meier (1967) 'Diurnal variations of the fattening response to prolactin in the golden topminnow, *Fundulus chrysotus*.' *J. Exp. Zool.*, *166*, 307-16.

Leggett, W.C. (1976) 'The American shad (*Alosa sapidissima*) with special reference to its migration and population dynamics in the Connecticut River' in D. Merriman and L.M. Thorpe (eds.), *The Connecticut River Ecological Study: The Impact of a Nuclear Power Plant*, *Amer. Fish. Soc. Monogr.*, *1*, 169-225.

Leggett, W.C. (1977) 'The ecology of fish migrations.' *Ann. Rev. Ecol. Syst.*, *8*, 285-308.

Leggett, W.C. and R.R. Whitney (1972) 'Water temperature and the migrations of American shad.' *US Fish. Wildl. Serv.*, *Fish. Bull.*, *70*, 659-70.

Leloup-Hatey, J. (1964) 'Corpuscules de Stannius et equilibre mineral chez l'Anquille (*Anguilla anguilla* L.).' *J. Physiol.*, *56*, 595-6.

LeRay, C., B. Bonnet, A. Febvre, F. Vallet and P. Pic (1970) 'Quelques activités périphériques des hormones thyroidiennes observeés chez *Mugil auratus* L. (Téléostéen Mugildé).' *Ann. Endocrinol.*, *31*, 567-72.

Lewis, U.J., R.N. Singh, B.K. Seavey, R. Lasker and G.E. Pickford (1972) 'Growth hormone and prolactinlike proteins of the shark (*Prionace glauca*).' *US Fish. Widl. Serv., Fish. Bull.*, *70*, 933-9.

Liley, N.R. (1969) 'Hormones and reproductive behavior in fishes' in W.S. Hoar and D.J. Randall (eds.), *Fish Physiology*, *Vol. III,* Academic Press, New York.

Lin, Y., G.H. Dobbs and A.L. DeVries (1974) 'Oxygen consumption and lipid content in red and white muscle of Antarctic fishes.' *J. Exp. Zool.*, *189*, 379-85.

Lindsey, C.C. and T.G. Northcote (1963) 'Life history of redside shiners, *Richardsonius balteatus,* with particular reference to movements in and out of Sixteen-mile Lake streams.' *J. Fish. Res. Board Can.*, *20*, 1001-30.

Lissmann, H.W. (1958) 'On the function and evolution of electric organs in fish.' *J. Exp. Biol.*, *35*, 156-91.

Liversage, R.A., B.W. Price, W.C. Clarke and D.G. Butler (1971) 'Plasma adrenocorticosteroid levels in adult *Fundulus heteroclitus* (killifish) following hypophysectomy and pectoral fin amputation.' *J. Exp. Zool.*, *178*, 23-8.

Loretz, C.A. and H.A. Bern (1981) 'Stimulation of sodium transport across the teleost urinary bladder by urotensin II.' *Gen. Comp. Endocrinol.*, *43*, 325-30.

Loretz, C.A., H.A. Bern, J.K. Foskett and J.R. Mainoya (1982) 'The caudal neurosecretory system and osmoregulation in fish' in D.S. Farmer and K. Lederis (eds.), *Neurosecretion: Molecules, Cells, Systems*, Plenum Publ. Corp., New York.

Love, R.M. (1958) 'Studies on the North Sea cod. III. Effects of starvation.' *J. Sci. Fd. Agric.*, *9*, 617-20.

Love, R.M. (1970) *The Chemical Biology of Fishes*, Academic Press, New York.

Love, R.M., I. Robertson, J. Lavetz and J.L. Smith (1974) 'Some biochemical characteristics of cod (*Gadus Morhua* L.) from the Faroe Bank compared with those from other fishing grounds.' *Comp. Biochem. Physiol.*, *47B*, 149-61.

Lovern, J.A. (1934) 'Fat metabolism in fishes. V. The fat of the salmon in its young freshwater stages.' *Biochem.*, *28*, 1961-3.

Loyacano, H.A., J.A. Chappell and S.A. Gauthreaux (1977) 'Sun-compass orientation in juvenile largemouth bass, *Micropterus salmoides*.' *Trans. Am. Fish. Soc.*, *106*, 77-9.

Ma, S.W.Y., Y. Shami, H.H. Messer and D.H. Copp (1974) 'Properties of Ca^{++}–ATPase from the gill of rainbow trout (*Salmo gairdneri*).' *Biochem. Biophys. Acta.*, *345*, 243-51.

Machemer, L. and K. Fiedler, (1965) 'Zur hormonalen Steurung des Schaumnestbaues beim Paradies-fish, *Macropedus opercularis* L. (Anabantidae, Teleostei).' *Naturwissenschaften*, *52*, 648.

MacLean, J.A. and J.H. Gee (1971) 'The effects of temperature on movements of prespawning brook sticklebacks, *Culaea inconstans*, in The Roseau River, Manitoba.' *J. Fish. Res. Board Can.*, *28*, 919-23.

Madison, D.M., R.M. Horrall, A.B. Stasko and A.D. Hasler (1972) 'Migratory movements of sockeye salmon (*Oncorhynchus nerka*) in coastal British Columbia as revealed by ultrasonic tracking.' *J. Fish. Res. Board Can.*, *29*, 1025-33.

Maetz, J. (1973) 'Na^+/NH_4^+, Na^+/H^+ exchanges and NH_3 movements across gill of *Carassius auratus.*' *J. Exp. Biol.*, *58*, 255-75.

Mainoya, J.R. and H.A. Bern (1982) 'Effects of teleost absorption of water and NaCl in Tilapia, *Sarotherodon mossambicus*, adapted to fresh water or seawater.' *Gen. Comp. Endocrinol.*, *47*, 54-8.

Maksimov, V.A. (1972) 'Some data on the ecology of the Kamchatka trout (*Salmo mykiss* Walbaum) from the Utkholok River.' *Vop. Ikhtiol.*, *12*, 759-65.

Malikova, E.M. (1959) 'Biochemical analyses of young salmon at the time of their transformation to a condition close to the smolt stage and during retention of smolts in fresh water.' *Fish. Res. Board Can. Trans. Ser. No. 232,* 1-19.

Markert, J.R., D.A. Higgs, H.M. Dye and D.W. MacQuarrie (1977) 'Influence of bovine growth hormone on growth rate, appetite, and food conversion of yearling coho salmon (*Oncorhynchus kisutch*) fed two diets of different composition.' *Can. J. Zool., 55,* 74-83.

Marshall, W.S. and H.A. Bern (1979a) 'Teleostean urophysis: urotensin II and ion transport across the isolated skin of a marine teleost.' *Science*, *204*, 519-21.

Marshall, W.S. and H.A. Bern (1979b) 'Active chloride transport by isolated fish skin: inhibition by the caudal neurohormone urotensin.' *Fed. Proc.*, *38*, 1057.

Marshall, W.S. and H.A. Bern (1981) 'Active chloride transport by the skin of a marine teleost is stimulated by urotensin I and inhibited by urotensin II.' *Gen. Comp. Endocrinol.*, *43*, 484-91.

Mason, J.C. (1975) 'Seaward movement of juvenile fishes, including lunar periodicity in the salmon fry.' *J. Fish. Res. Board Can.*, *23*, 63-91.

Mason, J.C. (975) 'Seaward movement of juvenile fishes, including lunar periodicity in the movement of coho salmon (*Oncorhynchus kisutch*) fry.' *J. Fish. Res. Board Can.*, *32*, 2542-7.

Masur, L., G. Burford, G. Moore and K. Lederis (1976) 'Isolation of urophysial hormone-binding proteins.' *Proc. Can. Fed. Biol. Sci.*, *19*, 24.

Matty, A.J. (1962) 'Effects of mammalian growth hormone on *Cottus scorpius* blood.' *Nature*, *195*, 506-7.

Matty, A.J. and M.J. Sheltawy (1967) 'The relation of thyroxine to skin purines in *Salmo irideus*.' *Gen. Comp. Endocrinol.*, *9*, 473.

Mayer, N., J. Maetz, D.K.O. Chan, M. Forster and I. Chester-Jones (1967) 'Cortisol, a sodium excreting factor in the eel (*Anguilla anguilla* L.) adapted to sea water.' *Nature*, *214*, 1118-20.

Mazeaud, F. (1973) Cited in M. Fontaine (1975) 'Physiological mechanisms in the migration of marine and amphhaline fish.' *Adv. Mar. Biol.*, *13*, 241-355.

McBride, J.R., R.A. MacLeod and D.R. Idler (1960) 'Seasonal variation in the collagen content of Pacific herring tissues.' *J. Fish. Res. Board Can.*, *17*, 913-18.

McCarthy, L.S. and A.H. Houston (1977) 'Na^+: K^+– and HCO_3–stimulated ATPase activities in the gills and kidneys of thermally acclimated rainbow trout, *Salmo gairdneri*.' *Can. J. Zool.*, *55*, 704-12.

McCleave, J. (1967) 'Homing and orientation of cutthroat trout (*Salmo clarki*) in Yellowstone Lake, with special reference to olfaction and vision.' *J. Fish. Res. Board Can.*, *24*, 2011-44.

McCleave, J.D. (1978) 'Rhythmic aspects of estuarine migration of hatchery-reared Atlantic salmon (*Salmo salar*) smolts.' *J. Fish. Biol.*, *12*, 559-70.

McCleave, J.D. and G.W. LaBar (1972) 'Further ultrasonic tracking and tagging studies of homing cutthroat trout (*Salmo clarki*) in Yellowstone Lake.' *Trans. Amer. Fish. Soc.*, *101*, 44-54.

McCleave, J.D. and J.H. Power (1978) 'Influence of weak electric and magnetic fields on turning behavior in elvers of the American eel *Anguilla rostrata*.' *Mar. Biol.*, *40*, 29-34.

McCleave, J.D., S.A. Rommel and C.L. Cathcart (1971) 'Weak electric and magnetic fields in fish orientation.' *Ann. NY Acad. Sci.*, *188*, 270-82.

McInerney, J.E. (1961) 'An experimental study of salinity preference and related migratory behavior of five Pacific salmon.' MSc thesis, University of British Columbia, Vancouver, Canada.

McInerney, J.E. (1964) 'Salinity preference: an orientation mechanism in salmon migration.' *J. Fish. Res. Board Can.*, *21*, 995-1018.

McKeown, B.A. (1970) 'Immunological identification and quantification of pituitary hormones in the sockeye salmon (*Oncorhynchus nerka*, Walbaum).' PhD thesis, Simon Fraser University, Burnaby, Canada.
McKeown, B.A. (1972) 'Effect of 2-Br-alpha-ergocryptine on fresh water survival in the teleosts, *Tiphophorus hellerii* and *Poecilia latipinna*.' *Experientia*, *28*, 675-6.
McKeown, B.A. and K.J. Brewer (1978) 'Control of prolactin secretion in teleosts with special reference to recent studies of Pacific salmon' in P.J. Gaillard and H.H. Boer (eds.), *Comparative Endocrinology*, Elsevier/North Holland Biomedical Press, Amsterdam.
McKeown, B.A. and C.A. Hazlett (1975) 'The uptake of tritiated leucine by the prolactin cell follicles of the coho salmon, *Oncorhynchus kisutch*.' *Can. J. Zool.*, *53*, 1195-200.
McKeown, B.A. and R.E. Peter (1976) 'The effect of photoperiod and temperature on the release of prolactin from the pituitary gland of the goldfish, *Carassius auratus* L.' *Can. J. Zool.*, *54*, 1960-8.
McKeown, B.A. and A.P. van Overbeeke (1969) 'Immunohistochemical localization of ACTH and prolactin in the pituitary gland of adult migratory sockeye salmon (*Oncorhynchus nerka*).' *J. Fish. Res. Board Can.*, *26*, 1837-46.
McKeown, B.A. and A.P. van Overbeeke (1972) 'Prolactin and growth hormone concentrations in the serum and pituitary gland of adult migratory sockeye salmon (*Oncorhynchus nerka*).' *J. Fish. Res. Board Can.*, *29*, 303-9.
McKeown, B.A., T.M. John and J.C. George (1976) 'Effect of vasotocin on plasma GH, free fatty acids and glucose in coho salmon (*Oncorhynchus kisutch*).' *Endocrinol. Experimentalis*, *10*, 45-51.
McKeown, B.A., J.F. Leatherland and T.M. John (1975) 'The effect of growth hormone and prolactin on the mobilization of free fatty acids and glucose in the kokanee salmon, *Oncorhynchus nerka*.' *Comp. Biochem. Physiol.*, *50B*, 425-30.
McLaren, I.A. (1963) 'Effects of temperatures on growth of zooplankton and the adaptive value of vertical migration.' *J. Fish. Res. Board Can.*, *20*, 685-727.
McLeay, D.J. (1975) 'Variations in the pituitary-interrenal axis and the abundance of circulating blood-cell types in juvenile coho salmon (*Oncorhynchus kisutch*), during stream residence.' *Can. J. Zool.*, *53*, 1882-91.
McNicol, R.E. and D.L. Noakes (1979) 'Caudal fin branding fish for individual recognition in behavior studies.' *Beh. Res. Meth. Instr.*, *11*, 95-7.
McPhail, J.D. and C.C. Lindsey (1970) 'Freshwater fishes of northwestern Canada and Alaska.' *Bull. Fish. Res. Board Can.*, *173*, 1-381.
Mehrle, P.M. and W.R. Fleming (1970) 'The effect of early and midday prolactin injection on the lipid content of *Fundulus kansae* held on a constant photoperiod.' *Comp. Biochem. Physiol.*, *36*, 597-603.
Meier, A.H. (1969) 'Diurnal variations of metabolic responses to prolactin in lower vertebrates.' *Gen. Comp. Endocrinol. Suppl.*, *2*, 55-62.
Meier, A.H. and A.J. Fivizzani (1980) 'Physiology of migration' in S.A. Gauthreaux, Jr (ed.), *Animal Migration, Orientation and Navigation*, Academic Press, New York.
Meier, A.H., T.N. Trobec, M.M. Joseph and T.M. John (1971) 'Temporal synergism of prolactin and adrenal steroids in the regulation of fat stores.' *Proc. Soc. Exp. Biol. Med.*, *137*, 408-15.
Mighell, J.L. (1969) 'Rapid cold-branding of salmon and trout with liquid nitrogen.' *J. Fish. Res. Board Can.*, *26*, 2765-9.
Miles, S.G. (1968) 'Laboratory experiments on the orientation of the adult American eel, *Anguilla rostrata*.' *J. Fish. Res. Board Can.*, *25*, 2143-55.
Milne, R.S. and J.F. Leatherland (1978) 'Effect of ovine TSH, thiourea, ovine prolactin and bovine growth hormone on plasma thyroxine and tri-idothyronine levels in rainbow trout, *Salmo gairdneri*.' *J. Comp. Physiol.*, *124*, 105-10.
Molenda, V.E. and K. Fiedler (1971) 'Die Wirkung von Prolaktin auf das Verhalten von Stichlings (*Gasterosteus aculeatus* L.).' *Z. Tierpsychol.*, *28*, 463.
Mommsen, T.P., C.J. French and P.W. Hochachka (1980) 'Sites and patterns of protein and amino acid utilization during the spawning migration of salmon.' *Can. J. Zool.*, *58*, 1785-99.
Moon, T.W. (1983) 'Metabolic reserves and enzyme activities with food deprivation in

immature American eels, *Anguilla rostrata* (LeSueur).' *Can. J. Zool., 61,* 802-11.
Muir, B.S. and A. Niimi (1972) 'Oxygen consumption of the euryhaline fish aholehole (*Kuhlia sandvicensis*) with reference to salinity, swimming, and food consumption.' *J. Fish. Res. Board Can.*, *29*, 67-77.
Muller, E.E. (1974) 'Growth hormone and the regulation of metabolism' in S.M. McCann (ed.), *Endocrine Physiology*, MTP International Review of Science, Physiology Series One, Vol. 5, pp. 141-78, Butterworth, London.
Mulligan, T.J., L. Lapi, R. Kieser, S.B. Yamada and D.L. Duewer (1983) 'Salmon stock identification based on elemental composition of vertebrae.' *Can. J. Fish. Aquat. Sci.*, *40*, 215-29.
Munz, F.W. and R.T. Swanson (1965) 'Thyroxine-induced changes in the proportions of visual pigments.' *Am. Zool.*, *5*, 683.
Murat, J.C., and A. Serfaty (1971) 'Variations saisonnières de l'effet de la thyroxine sur le métabolisme glucidique de la carpe.' *J. Physiol.* (Paris), *63*, 80-1.
Myers, G.S. (1949) 'Usage of anadromous, catadromous, and allied terms for migratory fishes.' *Copeia*, 89-97.
Nagahama, Y. (1973) 'Histo-physiological studies on the pituitary glands of some teleost fishes, with special reference to the classification of hormone-producing cells in the adenohypophysis.' *Mem. Fac. Fish. Hokkaido Univ.*, *21*, 1-63.
Nagahama, Y., B.A. Doneen, R.S. Nishioka and H.A. Bern (1973) 'Ultrastructure of the urinary bladder of the teleost *Gillichthys mirabilis* adapted to different salinities and treated with prolactin.' *Am. Zool.*, *13*, 1278.
Nagahama, Y., H. Kagawa and G. Young (1982) 'Cellular sources of sex steroids in teleost gonads.' *Can. J. Fish. Aqua. Sci.*, *39*, 56-64.
Nagahama, Y., R. Nishioka and H.A. Bern (1974) 'Structure and function of the transplanted pituitary in the seawater goby, *Gillichthys mirabilis*. I. The rostral pars distalis.' *Gen. Comp. Endocrinol.*, *22*, 21-34.
Nagai, M. and S. Ikeda (1973) 'Carbohydrate metabolism in fish. IV. Effect of dietary composition on metabolism of acetate-U-^{14}C and L-alanine-U-^{14}C in carp.' *Nippon Suisan Gakkaishi*, *39*, 633-43.
Naito, K. and F.H. Wilt (1962) 'The conversion of vitamin A[1] to retinene 2 in a freshwater fish.' *J. Biol. Chem.*, *237*, 3060-4.
Nakashima, B.S. and W.C. Leggett (1975) 'Yellow perch (*Perca flavescens*) biomass responses to different levels of phytoplankton and benthic biomass in Lake Memphremagog, Quebec-Vermont.' *J. Fish. Res. Board Can.*, *32*, 1785-97.
Narayansingh, T. and J.G. Eales (1975) 'The influence of physiological doses of thyroxine on the lipid reserves of starved and fed brook trout, *Salvelinus fontinalis* (Mitchill).' *Comp. Biochem. Physiol.*, *52B*, 407-12.
Neilson, J.D. (1983) 'Formation of otolith growth increments and their potential for assessing the early life history of chinook salmon (*Oncorhynchus tshawytscha*).' PhD thesis, Simon Fraser University, Burnaby, Canada.
Neilson, J.D. and G.H. Geen (1982) 'Otoliths of chinook salmon (*Oncorhynchus tshawytscha*): Daily growth increments and factors influencing their production.' *Can. J. Fish. Aquat. Sci.*, *39*, 1340-7.
Ng, T.B. and D.R. Idler (1978a) '"Big" and "little" forms of plaice vitellogenic and maturational hormones.' *Gen. Comp. Endocrinol.*, *34*, 408-20.
Ng, T.B. and D.R. Idler (1978b) 'A vitellogenic hormone with a large and small form from salmon pituitaries.' *Gen. Comp. Endocrinol.*, *35*, 189-95.
Noble, G.K., K.F. Kumpf and V.N. Billings (1938) 'The induction of brooding behavior in the jewel fish.' *Endocrinology*, *23*, 353.
Nordeng, H. (1971) 'Is the local orientation of anadromous fishes determined by pheromones?' *Nature*, *233*, 411-13.
Norris, D.O. (1966) 'Radiothyroidectomy in the salmonid fishes *Salmo gairdneri* Richardson and *Oncorhynchus tshawytscha* Walbaum.' PhD thesis, University of Washington.
Norris, D.O. (1969) 'Depression of growth following radiothyroidectomy of larval chinook salmon and steelhead trout.' *Trans. Am. Fish. Soc.*, *98*, 104-6.
Norris, D.O. (1980) *Vertebrate Endocrinology*, Lea and Febiger, Philadelphia.
Northcote, T.G. (1978) 'Migratory strategies and production in freshwater fishes' in S.D.

Gerking (ed.), *Ecology of Freshwater Fish Production*, John Wiley and Sons, New York.

Northcote, T.G. and B.W. Kelso (1981) 'Differential response to water current by two homozygous LDH phenotypes of young rainbow trout (*Salmo gairdneri*).' *Can. J. Fish. Aquat. Sci.*, *38*, 348-52.

Nozaki, M. and H. Kobayashi (1979) 'Distribution of LHRH-like substance in the vertebrate brain as revealed by immunohistochemistry.' *Arch. Histol. Jpn.*, *42*, 201-19.

Ogawa, M. (1974) 'The effect of bovine prolactin, sea water and environmental calcium on water influx in isolated gills of the euryhaline teleosts *Anguilla japonica* and *Salmo gairdneri*.' *Comp. Biochem. Physiol.*, *49A*, 545-53.

Ogawa, M. (1975) 'The effects of prolactin, cortisol and calcium-free environment on water influx in isolated gills of Japanese eels, *Anguilla japonica*.' *Comp. Biochem. Physiol.*, *52A*, 539-43.

Ogawa, M. (1977) 'The effect of hypophysectomy and prolactin treatment on the osmotic water influx into the isolated gills of the Japanese eel (*Anguilla japonica*).' *Can. J. Zool.*, *55*, 872-6.

Oguri, M. (1960) 'Studies on the adrenal glands of teleosts. IV. Histochemistry of the interrenal cells of fishes.' *Bull. Jap. Soc. Scien. Fish.*, *26*, 447-51.

Olivereau, M. (1954) 'Hypophyse et glande thyroide chez les poissons. Etude histophysiologique de quelques correlations endocriniennes, en particular chez *Salmo salar* L.' *Ann. Inst. Oceanogr.*, *29*, 95-296.

Olivereau, M. (1962) 'Modifications de l'interrénal du smolt (*Salmo salar* L.) au cours du passage d'eau douce en eau de mer.' *Gen. Comp. Endocrinol.*, *2*, 565-73.

Olivereau, M. (1971) 'Effect of reserpine on the eel. I. Prolactin secreting cells in the pituitary gland of male eels.' *Z. Zellforsch. Mikrosk. Anat.*, *121*, 232-43.

Olivereau, M. (1975) 'Histophysiologie de l'axe hypophys-corticosurrenalien chez le saumon de l'Atlantique (cycle en eau douce, vie thalassique et reproduction).' *Gen. Comp. Endocrinol*, *27*, 9-27.

Oshima, K. and A. Gorbman (1966a) 'Olfactory responses in the forebrain of goldfish and their modification by thyroxine treatment.' *Gen. Comp. Endocrinol.*, *7*, 398-409.

Oshima, K. and A. Gorbman (1966b) 'Influence of thyroxine and steroid hormones on spontaneous and evoked unitary activity in the olfactory bulb of goldfish.' *Gen. Comp. Endocrinol.*, *7*, 482-91.

Oshima, K. and A. Gorbman (1968) 'Modification by sex hormones of the spontaneous and evoked bulbar electrical activity in goldfish.' *J. Endocrinol.*, *40*, 409-20.

Oshima, K., A. Gorbman and H. Shimada (1969) 'Memory-blocking agents: effects on olfactory discrimination in homing salmon.' *Science*, *165*, 86-8.

Oshima, K., W.E. Hahn and A. Gorbman (1969) 'Olfactory discrimination of natural waters by salmon.' *J. Fish. Res. Board Can.*, *26*, 2111-21.

Osterdahl, L. (1969) 'The smolt run of a small Swedish river' in T.G. Northcote (ed.), *Symposium on Salmon and Trout in Streams*, MacMillan Lectures in Fisheries, Univ. of British Columbia, Vancouver.

Otto, R.G. and J.E. McInerney (1970) 'Development of salinity preference in presmolt coho salmon, *Oncorhynchus kisutch*.' *J. Fish. Res. Board Can.*, *27*, 793-800.

Ovchinnikov, V.V., S.I. Gleizer and G.Z. Galaktionov (1973) 'Features of orientation of the European eel (*Anguilla anguilla* L.) at some stages of migration.' *J. Ichthyol.*, *13*, 455-63.

Pals, N. and A.A.C. Schoenhage (1979) 'Marine electric fields and fish orientation.' *J. Physiol.*, *75*, 349-53.

Pang, P.K.T. (1973) 'Endocrine control of calcium metabolism in teleosts.' *Amer. Zool.*, *13*, 775-92.

Pang, P.K.T., R.K. Pang and W.H. Sawyer (1974) 'Environmental calcium and the sensitivity of killifish (*Fundulus heteroclitus*) in bioassays for the hypocalcemic response to Stannius corpuscles from killifish and cod (*Gadus morhua*).' *Comp. Biochem. Physiol.*, *94*, 548-55.

Pandy, S. and J.F. Leatherland (1970) 'Comparison of the effects of methallibure and thiourea on the testis, thyroid, and the adenohypophysis of the adult and juvenile guppy, *Poecilia reticulata* Peters.' *Can. J. Zool.*, *48*, 445-50.

Pannella, G. (1971) 'Fish otoliths: daily growth layers and periodical patterns.' *Science*, *173*, 1124-6.
Papi, F. and L. Pardi (1963) 'On the lunar orientation of sandhoppers (*Amphipoda talitridae*).' *Biol. Bull.*, *124*, 97-105.
Pardo, R.J. and V.L. de Vlaming (1976) '*In vivo* and *in vitro* effects of prolactin on lipid metabolism in the cyprinid teleost *Notemigonus crysoleucas*.' *Copeia*, pp. 563-73.
Parsons, J.A. (1979) 'Physiology of parathyroid hormone' in L.J. DeGroot (ed.), *Endocrinology*, Grune and Stratton, New York.
Patent, G.J. and P.P. Foa (1971) 'Radioimmunoassay of insulin in fishes, experiments *in vivo* and *in vitro*.' *Gen. Comp. Endocrinol.*, *16*, 41-6.
Patten, B.C. (1964) 'The rational decision process in salmon migration.' *J. Cons. Cons. Perm. Int. Explor. Mer.*, *28*, 410-17.
Pavshtiks, E.A. (1959) 'Seasonal changes in plankton and feeding migrations of herring' in *The herring of the North European Basin and adjacent seas. US Fish Wildl. Serv. Spec. Sci. Rep. Fish.*, *327*, 104-39.
Payan, P. and J. Maetz (1973) 'Branchial sodium transport mechanisms in *Scyhiorhinus canicula*: Evidence for Na^+/NH_4^+ and Na^+/H^+ exchanges and a role for carbonic anhydrase.' *J. Exp. Biol.*, *58*, 487-502.
Pearson, D., J.E. Shively, B.R. Clarke, I.I. Geshwind, M. Barkley, R.S. Nishioka and H.A. Bern (1980) 'Urotensin II: A somatostatin-like peptide in the caudal neurosecretory system of fishes.' *Proc. Nat. Acad. Sci. USA*, *77*, 5021-4.
Peter, R.E. (1979) 'The brain and feeding behavior' in W.S. Hoar, D.J. Randall and J.R. Brett (eds.), *Fish Physiology*, *Vol. VIII*, Academic Press, New York.
Peter, R.E. (1982) 'Neuroendocrine control of reproduction in teleosts.' *Can. J. Fish. Aqua. Sci.*, *39*, 48-55.
Peter, R.E., J.P. Chang, C.S. Nahorniak and L.W. Crim (1982) 'Pars distalis transplants in the goldfish: additional evidence for gonadotropin release-inhibitory hormone.' *Western Regional Conf. on Comp. Endocrinol.*, Seattle.
Peter, R.E. and L.W. Crim (1978) 'Hypothalamic lesions of goldfish: effects on gonadal recrudescence and gonadotropin secretion.' *Ann. Biol. Anim. Bioch. Biophys.*, *18*, 819-23.
Peter, R.E. and L.W. Crim (1979) 'Reproductive endocrinology of fishes: Gonadal cycles and gonado-tropin in teleosts.' *Ann. Rev. Physiol.*, *41*, 323-35.
Peter, R.E. and C.R. Paulencu (1980) 'Involvement of the preoptic region in goldfish, *Carassius auratus*.' *Neuroendocrinology*, *31*, 133-41.
Peter, R.E., L.W. Crim, H.J.Th. Goos and J.W. Crim (1978) 'Lesioning studies on the gravid female goldfish: neuroendocrine regulation of ovulation.' *Gen. Comp. Endocrinol.*, *35*, 391-401.
Peter, R.E., O. Kah, C.R. Paulencu, H. Cook and A.L. Kyle (1980) 'Brain lesions and short-term endocrine effects of monosodium L-glutamate in goldfish, *Carassius auratus*.' *Cell. Tissue Res.*, *212*, 429-42.
Peter, R.E., E.A. Monckton, and B.A. McKeown (1976) 'The effects of gold thioglucose on food intake, growth, and forebrain histology in goldfish, *Carassius auratus*.' *Physiol. Behav.*, *17*, 303-12.
Phinney, D.E., D.M. Miller and M.L. Dahlberg (1967) 'Mass-marking young salmonids with fluorescent pigment.' *Trans. Am. Fish. Soc.*, *96*, 157-62.
Pic, P., N. Mayer-Gostan and J. Maetz (1973) 'Sea-water teleosts: presence of α- and β-adrenergic receptors in the gill regulating salt extrusion and water permeability' in L. Bolis, K. Schmidt-Neilson and S.H.P. Moddrell (eds.), *Comparative Physiology*, Elsevier/North Holland Biomedical Press, Amsterdam.
Pickford, G.E. (1953) 'The response of hypophysectomized male *Fundulus* to injections of purified beef growth hormone.' *Bull. Bingham Oceanogr. Collect.*, *14*, 46-68.
Pickford, G.E. (1954) 'The response of hypophysectomized male killifish to purified fish growth hormone, as compared with the response to purified beef growth hormone.' *Endocrinology*, *55*, 274-87.
Pickford, G.E. (1957) 'The growth hormone' in G.E. Pickford and J.W. Atz (eds.), *The Physiology of the Pituitary Gland of Fishes*, NY Zool. Soc., New York, Part IV, pp.84-99.

Pickford, G.E. (1959) 'The nature and physiology of pituitary hormones of fishes' in A. Gorbman (ed.), *Comparative Endocrinology*, John Wiley, New York, pp. 404-20.
Pickford, G.E. (1973) 'Introductory remarks.' *Am. Zool.*, *13*, 711-17.
Pickford, G.E. and J.G. Phillips (1959) 'Prolactin, a factor in promoting survival of hypophysectomised killifish in fresh water.' *Science*, *130*, 454-5.
Pickford, G.E., R.W. Griffith, J. Torretti, E. Hendley and F.H. Epstein (1970) 'Branchial reduction and renal stimulation of (Na^+, K^+)–ATPase by prolactin in hypophysectomized killifish in fresh water.' *Nature*, *228*, 378-9.
Pickford, G.E., Wilhelmi, A.E., and Nussbaum, N. (1959) 'Comparative studies of the response of hypophysectomized killifish, *Fundulus heteroclitus*, to growth hormone preparations.' *Anat. Rec.*, *134*, 624-5.
Piggins, D.J. (1962) 'Thyroid feeding of salmon parr.' *Nature*, *195*, 1017-18.
Plack, P.A. and P.M. J. Woodhead (1966) 'Vitamin A compounds and lipids in blood of cod, *Gadus morhua* from the Arctic in relation to gonadal maturation.' *J. Mar. Biol. Ass. UK*, *46*, 547-59.
Quinn, T.P. (1980) 'Evidence for celestial and magnetic compass orientation in lake migrating sockeye salmon fry.' *J. Comp. Physiol.*, *137*, 243-8.
Quinn, T.P. (1982a) 'Intra-specific differences in sockeye salmon by compass orientation mechanisms.' *Proc. Salmon and Trout Migratory Behavior Symp.*, Seattle, pp. 79-85.
Quinn, T.P. (1982b) 'A model for salmon navigation on the high seas.' *Proc. Salmon and Trout Migratory Behavior Symp.*, pp. 229-37.
Quinn, T.P. and E.L. Brannon (1982) 'The use of celestial and magnetic cues by orienting sockeye salmon smolts.' *J. Comp. Physiol.*, *147*, 547-52.
Quinn, T.P., R.T. Merrill and E.L. Brannon (1981) 'Magnetic field detection in sockeye salmon.' *J. Exp. Zool.*, *217*, 137-42.
Rayleigh, R.F. (1971) 'Innate control of migration of salmon and trout fry from natal gravels to rearing areas.' *Ecology*, *52*, 291-7.
Rayleigh, R.F. and D.W. Chapman (1971) 'Genetic control in lakeward migration of cutthroat trout fry.' *Trans. Am. Fish. Soc.*, *100*, 33-40.
Rayleigh, R.F. (1967) 'Genetic control in the lakeward migrations of sockeye salmon (*Oncorhynchus nerka*) fry.' *J. Fish. Res. Board Can.*, *24*, 2613-22.
Rao, G.M.M. (1968) 'Oxygen consumption of rainbow trout (*Salmo gairdneri*) in relation to activity and salinity.' *Can. J. Zool.*, *46*, 781-6.
Rappaport, D.A. and H.F. Daginawala (1968) 'Changes in nuclear RNA of brain induced by olfaction in catfish.' *J. Neurochem.*, *15*, 991-1006.
Refstie, T. (1982) 'The effect of feeding thyroid hormones on salt water tolerance and growth rate of Atlantic salmon.' *Can. J. Zool.*, *60*, 2706-12.
Richkus, W.A. (1974) 'Factors influencing the seasonal and daily patterns of alewife (*Alosa pseudoharengus*) migration in a Rhode Island river.' *J. Fish. Res. Board Can.*, *31*, 1485-97.
Richman, N.H. and E.B. Barnawell (1978) 'Ouabain urophyseal extract on sodium transport on the skin and intestine of channel catfish, *Ictalurus punctatus*.' *Am. Zool.*, *18*, 651.
Ricker, W.E. (1979) 'Growth rates and models' in W.S. Hoar, D.J. Randall and J.R. Brett (eds.), *Fish Physiology*, *Vol. VIII*, Academic Press, New York.
Robertson, O.H. (1955) 'Science, salmon and trout.' *Trans. Ass. Am. Physns.*, *68*, 33-41.
Robertson, O.H. and B.C. Wexler (1959) 'Hyperplasia of the adrenal cortical tissue in Pacific salmon (genus *Oncorhynchus*) and rainbow trout (*Salmo gairdneri*) accompanying sexual maturation and spawning.' *Endocrinology*, *65*, 225-38.
Robertson, O.H., M.A. Krapp, C.B. Favour, S. Hane and S.F. Thomas (1961a) 'Physiological changes occurring in the blood of the Pacific salmon (*Oncorhynchus tshawytscha*).' *Endocrinology*, *68*, 733-46.
Robertson, O.H., M.A. Krupp, S.F. Thomas, C.B. Favour, S. Hane and B.C. Wexler (1961b) 'Hyperadrenocorticism in spawning migrating and non-migrating rainbow trout (*Salmo gairdneri*): Comparison with Pacific salmon (genus *Oncorhynchus*).' *Gen. Comp. Endocrinol.*, *1*, 473-84.
Rommel, S.A. and J.D. McCleave (1972) 'Oceanic electric fields: perception by American eels.' *Science*, *176*, 1233-5.

Royce, W.F., L.S. Smith and A. Hartt (1968) 'Models of oceanic migrations of Pacific salmon and comments on guidance mechanisms.' *US Fish. Wildl. Serv., Fish. Bull., 66,* 441-62.

Rozin, P. and Mayer, J. (1961) 'Regulation of food intake in the goldfish.' *Am. J. Physiol., 201,* 968-74.

Sage, M. (1967) 'Responses of pituitary cells of *Poecilia* to changes in growth induced by thyroxine and thiourea.' *Gen. Comp. Endocrinol., 8,* 314-19.

Sage, M. (1968) 'Respiratory and behavioral responses of *Poecilia* to treatment with thyroxine and thiourea.' *Gen. Comp. Endocrinol., 10,* 304-9.

Saila, S.B. (1961) 'A study of winter flounder movements.' *Limnol. Oceanogr., 6,* 292-8.

Saila, S.B. and R.A. Shappy (1963) 'Random movement and orientation in salmon migration.' *J. Cons. Cons. Perm. Int. Explor. Mer., 28,* 153-66.

Sakaguchi, M., M. Sugiyama, T. Sugiyama and A. Kawai (1970) 'Histidine metabolism in fish. IV. Comparative study on histidine deaminases from muscle and liver of mackerel and from bacteria.' *Nippon Suisan Gokkaishi, 36,* 200-6.

Saunders, R.L. (1965) 'Adjustment of buoyancy in young Atlantic salmon and brook trout by changes in swim-bladder volume.' *J. Fish. Res. Board Can., 22,* 335-52.

Saunders, R.L. and E.B. Henderson (1970) 'Influence of photoperiod on smolt transformation and growth of Atlantic salmon (*Salmo salar*).' *J. Fish. Res. Board Can., 27,* 1295-311.

Sawyer, W.H. (1972) 'Neurohypophyseal hormones and water and sodium excretion in African lungfish.' *Gen. Comp. Endocrinol. Suppl., 3,* 345-9.

Schmidt, J. (1922) 'The breeding places of the eel.' *Phil. Trans. R. Soc. B., 211,* 179-208.

Scholz, A.T., R.M. Horrall, J.C. Cooper and A.D. Hasler (1976) 'Imprinting to chemical cues: The basis of homestream selection in salmon.' *Science, 192,* 1247-9.

Scholz, A.T., R.M. Horrall, J.C. Cooper, A.D. Hasler, D.M. Madison, R.J. Poff and R. Daly (1975) 'Artificial imprinting of salmon and trout in Lake Michigan.' *Wis. Dept. nat. Res. Fish. Mgmt. Rep., 80,* 1-45.

Scholz, A., D.M. Madison, A.B. Stasko, R.M. Horrall and A.D. Hasler (1972) 'Orientation of salmon in response to currents in or near the homestream.' *Am. Zool., 12,* 654.

Schreibman, M.P., L.R. Halpern, H.J.Th. Goos and H. Margolis-Kazan (1979) 'Identification of luteinizing hormone-releasing hormone (LH-RH) in the brain and pituitary gland of a fish by immunocytochemistry.' *J. Exp. Zool., 210,* 153-9.

Schreibman, M.P., J.F. Leatherland and B.A. McKeown (1973) 'The functional morphology of the teleost pituitary gland.' *Amer. Zool., 13,* 719-42.

Schwassmann, H.O. (1967) 'Orientation of Amazonian fishes to the equatorial sun.' *Atas do Simposio sobre a Biota Amazonica (Limnologia), 3,* 201-20.

Schwassmann, H.O. (1971) 'Biological rhythms' in W.S. Hoar and D.J. Randall (eds.), *Fish Physiology, Vol. VI,* Academic Press, New York.

Schwassmann, H.O. and A.D. Hasler (1964) 'The role of the sun's altitude in orientation of fish.' *Physiol. Zool., 37,* 163-78.

Scott, A.P., V.J. Bye and S.M. Baynes (1980) 'Seasonal variations in sex steroids of female rainbow trout (*Salmo gairdneri* Richardson).' *J. Fish. Biol., 17,* 587-92.

Scott, J.L. (1944) 'The effects of steroids on the skeleton of the poecilid fish *Lebistes reticulatus*.' *Zoologica* (NY), *29,* 49-53.

Secondat, M. (1950) 'Influence de l'exercise musculaire sur la valeur de glyceme de la carpe (*Cyprinus carpio* L.).' *CR. Acad. Sci., 231,* 796.

Selset, R. and K.B. Døving (1980) 'Behaviour of mature anadromous char (*Salmo alpinus* L.) towards odorants produced by smolts of their own population.' *Acta Physiol. Scand., 108,* 113-22.

Selye, H. (1950) *The Physiology and Pathology of Exposure to Stress,* Acta Inc. Medical Publishers, Montreal.

Sklower, A. (1930) 'Die Bedeutung der Schilddüse für die Metamorphose des Aales und der Plattfische.' *Forsch. Fortschr., 6,* 435-6.

Slaney, P.A. and T.G. Northcote (1974) 'Effects of prey abundance on density and territorial behavior of young rainbow trout (*Salmo gairdneri*) in laboratory stream channels.' *J. Fish. Res. Board Can., 31,* 1201-9.

Slaney, P.A. (1972) 'Effects of prey abundance on distribution, density and territorial behavior of young rainbow trout in laboratory stream channels.' *J. Fish. Res. Board Can.*, *31*, 1201-9.

Slaney, P.A. (1972) 'Effects of prey abundance on distribution, density and territorial behavior of young rainbow trout in streams.' MSc thesis, University of British Columbia, Vancouver, Canada.

Smith, D.C.W. (1956) 'The role of the endocrine organs in the salinity tolerance of the trout.' *Mem. Soc. Endocrinol.*, *5*, 83-101.

Smith, M.A.K. and A. Thorpe (1977) 'Endocrine effects on nitrogen excretion in the euryhaline teleost, *Salmo gairdneri*.' *Gen. Comp. Endocrinol.*, *32*, 400-6.

Smith, M.W. (1967) 'Influences of temperature acclimatization on the temperature-dependence and ouabain-sensitivity of goldfish intestinal adenosine triphosphatase.' *Biochem. J.*, *105*, 65-71.

So, Y.P. and J.C. Fenwick (1977) 'Relationship between net ^{45}Ca influx across perfused eel gill and the development of post-stanniectomy hypercalcemia.' *J. Exp. Zool.*, *200*, 259-64.

Sokabe, H., H. Nishimura, M. Ogawa and M. Oguri (1970) 'Determination of renin in the corpuscles of Stannius of the teleost.' *Gen. Comp. Endocrinol.*, *14*, 510-16.

Solomon, D.J. (1973) 'Evidence for pheromone-influenced homing by migrating Atlantic salmon, *Salmo salar* (L.).' *Nature*, *244*, 231-2.

Solomon, D.J. (1978) 'Migration of smolts of Atlantic salmon (*Salmo salar* L.) and sea trout (*Salmo trutta* L.) in a chalkstream.' *Env. Biol. Fish.*, *3*, 233-9.

Sotthibandhu, S. and R.R. Baker (1979) 'Celestial orientation by the large yellow underwing moth, *Noctua pronuba*.' *Anim. Behav.*, *27*, 786-800.

Speidel, C.C. (1919) 'Gland cells of internal secretion in the spinal cord of the skates.' *Carnegie Inst. Wash. Papers*, *13*, 1-31.

Speidel, C.C. (1922) 'Further comparative studies in other fishes of cells that are homologous to the large irregular gland cells in the spinal cord of skates.' *J. Comp. Neurol.*, *34*, 303-17.

Spieler, R.E. (1975) 'Circadian and circannual serum prolactin levels in some estuarine fishes: endocrinological, ecological, and maricultural implications.' PhD thesis, Louisiana State University.

Spierler, R.E. and A.H. Meier (1975) 'Short-term serum prolactin concentrations in goldfish (*Carassius auratus*) subjected to serial sampling and restraint.' *J. Fish. Res. Board Can.*, *33*, 183-6.

Spieler, R.E., A.H. Meier and H.C. Loesch (1976a) 'Seasonal variations in circadian levels of serum prolactin in striped mullet, *Mugil cephalus*.' *Gen. Comp. Endocrinol.*, *29*, 150-60.

Spieler, R.E., A.H. Meier and H.C. Loesch (1976b) 'Photoperiodic effects on salinity selection in the gulf killifish, *Fundulus grandis*.' *Copeia*, 605-8.

Spieler, R.E., A.H. Meier and T.A. Noeske (1978) 'Temperature-induced phase shift of daily rhythm of serum prolactin in gulf killifish.' *Nature*, *271*, 469-70.

Stacey, N.E., A.F. Cook and R.E. Peter (1979a) 'Spontaneous and gonadotropin-induced ovulation in the goldfish, *Carassius auratus* L.: effects of external factors.' *J. Fish. Biol.*, *15*, 349-61.

Stacey, N.E., A.F. Cook and R.E. Peter (1979b) 'Ovulatory surge of gonadotropin in the goldfish, *Carassius auratus*.' *Gen. Comp. Endocrinol.*, *37*, 246-9.

Stanley, J.G. (1974) 'Nitrogen and phosphorus balance of grass carp, *Ctenopharyngodon idella*, fed elodea, *Egeria densa*.' *Trans. Am. Fish. Soc.*, *103*, 587-92.

Stanley, J.G. and W.R. Fleming (1967) 'Effect of prolactin and ACTH on the serum and urine levels of *Fundulus kansae*.' *Comp. Biochem. Physiol.*, *20*, 199-208.

Stanley, L.L. and G.L. Tescher (1931) 'Activity of goldfish in testicular substance diet.' *Endocrinology*, *15*, 55-6.

Stasko, A.B., R.M. Horrall, A.D. Hasler and D. Stasko (1973) 'Coastal movements of mature Fraser River pink salmon (*Oncorhynchus gorbuscha*) as revealed by ultrasonic tracking.' *J. Fish. Res. Board Can.*, *30*, 1309-11.

Stasko, A.B., R.M. Horrall, and A.D. Hasler (1976) 'Coastal movements of adult Fraser River sockeye salmon (*Oncorhynchus nerka*) observed by ultrasonic tracking.' *Trans. Amer. Fish. Soc.*, *105*, 64-71.

Sterba, G. (1955) 'Das Adrenal und Interrenal system im Lebensablauf von *Petromyzon planeri* Block. I. Morphologie und Histologie einschliesslich Histogenese.' *Zool. Amz.*, *155*, 151-69.
Stevens, E.D. (1968) 'The effect of exercise on the distribution of blood to various organs in rainbow trout.' *Comp. Biochem. Physiol.*, *25*, 615.
Stewart, K.W. (1962) 'Observations on the morphology and optical properties of the adipose eyelid of fishes.' *J. Fish. Res. Board. Can.*, *19*, 1161-2.
Stickney, A.P. (1972) 'The locomotor activity of juvenile herring (*Clupea harengus harengus* L.) in response to changes in illumination.' *Ecology*, *53*, 438-45.
Stimpson, J.H. (1965) 'Comparative aspects of the control of glycogen utilization in vertebrate liver.' *Comp. Biochem. Physiol.*, *15*, 187-97.
Sundararaj, B.I. and S.V. Goswami (1965) '"Seminal vesicle" response of intact castrate and hypophysectomized catfish, *Heteropneustes fossilis* (Block) to testosterone propionate, prolactin, and growth hormone.' *Gen. Comp. Endocrinol.*, *5*, 464-74.
Sundararaj, B.I. and P. Keshavanath (1976) 'Effects of melatonin and prolactin treatment on the hypophysial-ovarian system in the catfish, *Heteropneustes fossilis* (Block).' *Gen. Comp. Endorcinol.*, *29*, 84-96.
Sundararaj, B.I. and S. Vasal (1976) 'Photoperiod and temperature control in the regulation of reproduction in the female catfish *Heteropneustes fossilis*.' *J. Fish. Res. Board Can.*, *33*, 959-73.
Suterlin, A.M. and N. Suterlin (1971) 'Electrical responses of the olfactory epithelium of Atlantic salmon (*Salmo salar*).' *J. Fish. Res. Board Can.*, *28*, 565-72.
Suterlin, A.M. and R. Gray (1973) 'Chemical basis for homing of Atlantic salmon (*Salmo salar*) to a hatchery.' *J. Fish. Res. Board Can.*, *30*, 985-9.
Svärdson, G. (1943) 'Studien über den Zusammenhang zwischen Geschlechtsreife und Wachstum bei *Lebistes*.' *Medd. Statens Undersök. Försöksanst. Sötvattenfisket*, *21*, 1-48.
Swift, D.R. (1964) 'Activity cycles in the brown trout (*Salmo trutta* L.) 2. Fish artificially fed.' *J. Fish. Res. Board Can.*, *21*, 133-8.
Swift, D.R. and G.E. Pickford (1962) 'Seasonal variations in the growth hormone content of pituitary glands of the perch, *Perca fluviatilis* L.' *Am. Zool.*, *2*, 451.
Swift, D.R. and G.E. Pickford (1965) 'Seasonal variations in the hormone content of the pituitary gland of the perch, *Perca fluviatilus* L.' *Gen. Comp. Endocrinol.*, *5*, 354-65.
Taranenko, N.F. (1966) 'K dinamike chislennosti azovskoi Khamsy.' *Tr. Azovo-Chernomorsk-Nauchno-Issled. Inst. Morsk. Rybn. Khoz. Okeanogr.*, *24*, 3-16.
Tashima, L. and G.F. Cahill (1968) 'Effects of insulin in the toadfish, *Opsanus tau*.' *Gen. Comp. Endocrinol.*, *11*, 262-71.
Tchernavin, V. (1939) 'The origin of salmon.' *Salmon and Trout Mag.*, *95*, 1-21.
Teichmann, H. (1962) 'Die Chemorezeption der Fische.' *Ergeb. Biol.*, *25*, 177-205.
Templeman, W. and G. Andrews (1956) 'Jellied condition in the American plaice *Hippoglossoides platessoides* (Fabricius).' *J. Fish. Res. Board Can.*, *13*, 147-82.
Tesch, F.-W. (1974) 'Influence of geomagnetism and salinity on the directional choice of eels.' *Helgänder wiss Meeresunters*, *26*, 382-95.
Thompson, M.H. and R.N. Farragut (1965) 'Depletion of contractile proteins and deposition of collagen during gonadal maturation in the herring, *Clupea pallasi*. *Fish. Ind. Res. 3*, 47-53.
Thornburn, C.C. and A.J. Matty (1963) 'The effect of thyroxine on some aspects of nitrogen metabolism in the goldfish (*Carrassius auratus*) and the trout (*Salmo trutta*).' *Comp. Biochem. Physiol.*, *8*, 1-12.
Thorpe, A. and B.W. Ince (1974) 'The effects of pancreatic hormones, catecholamines, and glucose loading on blood metabolites in the Northern pike (*Esox lucius* L.).' *Gen. Comp. Endocrinol.*, *23*, 29-44.
Tucker, D.W. (1959) 'A new solution to the Atlantic eel problem.' *Nature*, *183*, 495-501.
Turner, C.D. and J.T. Bagnara (1976) *General Endocrinology*, W.B. Saunders & Co., London.
Tyler, A.V., nd R.S. Dunn (1976) 'Ration, growth, and measures of somatic and organ condition in relation to meal frequency in winter flounder, *Pseudopleuronectes americanus*, with hypotheses regarding population homeostatis.' *J. Fish. Res. Board Can.*, *33*, 63-75.

Ueda, K., T.J. Hara and A. Gorbman (1967) 'Electroencephalographic studies on olfactory discrimination in adult spawning salmon.' *Comp. Biochem. Physiol.*, *21*, 133-43.

Ueda, K., T.J. Hara, M. Satou and S. Kaji (1971) 'Electrophysiological studies of olfactory discrimination of natural waters by hime salmon, a land-locked Pacific salmon, *Oncorhynchus nerka*.' *J. Fac. Sci. Tokyo Univ. Sec. IV*, *12*, 167-82.

Upadhyay, S.N. (1977) 'Morphology of immature gonads and experimental studies on the induction of gametogenesis in juvenile rainbow trout (*Salmo gairdneri* R.).' PhD thesis, L'Univ. Pierre et Marie Curie, Paris.

Urist, M.R. and A.R. Schjeide (1961) 'The partition of calcium and protein in the blood of oviparous vertebrates during estrus.' *J. Gen. Physiol.*, *44*, 743-56.

Utida, S., T. Hirano, H. Oide, M. Ando, D.W. Johnson and H.A. Bern (972) 'The role of the intestine and urinary bladder in teleost osmoregulation.' *Gen. Comp. Endocrinol. Suppl.*, *3*, 317-27.

Utida, S., M. Kamiya, D.W. Johnson and H.A. Bern (1974) 'Effects of freshwater adaptation and of prolactin on sodium-potassium-activated adenosine triphosphotase activity in the urinary bladder of two flounder species.' *J. Endocrinol.*, *62*, 11 14.

Van Overbeeke, A.P. and J.R. McBride (1971) 'Histological effects of 11-keto-testosterone, 17α-methyltestosterone, estradiol, estradiol cypionate, and cortisol on the interrenal tissue, thyroid gland and pituitary gland of gonadectomized sockeye salmon (*Oncorhynchus nerka*).' *J. Fish. Res. Board Can.*, *28*, 477-84.

Vanstone, W.E., E. Roberts and H. Tsuyaki (1964) 'Changes in the multiple hemoglobin patterns of some Pacific salmon, Genus *Oncorhynchus*, during the parr-smolt transformation.' *Can. J. Physiol. Pharmacol.*, *42*, 697-704.

Vladykov, V.D. (1964) 'Quest for the true breeding area of the American eel (*Anguilla rostrata* LeSueur).' *J. Fish. Res. Board Can.*, *21*, 1523-30.

Vodicnik, M.J., J. Olcese, G. Delahunty and V. de Vlaming (1979) 'The effects of blinding, pinealectomy and exposure to constant dark conditions on gonadal activity in the female goldfish, *Carassius auratus*.' *Environ. Biol. Fish.*, *4*, 173-7.

von Frisch, K. (1949) 'Die Polarisation des Himmelslichtes als orientierender Faktor bie den Tanzen der Bienen.' *Experientia*, *5*, 142-8.

Wagner, G.F. and B.A. McKeown (1981) 'Immunocytochemical localization of hormone-producing cells within the pancreatic islets of the rainbow trout (*Salmo gairdneri*).' *Cell. Tiss. Res.*, *221*, 181-92.

Wagner, G.F. and B.A. McKeown (1982) 'Changes in plasma insulin and carbohydrate metabolism of zinc-stressed rainbow trout, *Salmo gairdneri*.' *Can. J. Zool.*, *60*, 2079-84.

Wagner, G.F. and B.A. McKeown (1984) 'Isolation, purification and partial characterization of growth hormone from two species of Pacific salmon' in B. Lofts (ed.), *Comparative Endocrinology*, Hong Kong University Press, Hong Kong.

Wagner, G.F., J.C. Brown and B.A. McKeown (1984) 'Coho salmon (*Oncorhynchus kisutch*) growth hormone: Purification, characterization and comparative bioassays.' *Gen. Comp. Endocrinol*. (In review).

Wagner, H.H. (1969) 'Effect of stocking location of juvenile steelhead trout (*Salmo gairdneri*) on adult catch.' *Trans. Am. Fish. Soc.*, *98*, 27-34.

Wagner, H.H. (1974a) 'Photoperiod and temperature regulation of smoltification in steelhead trout (*Salmo gairdneri*).' *Can. J. Zool.*, *52*, 219-34.

Wagner, H.H. (1974b) 'Seawater adaptation independent of photoperiod in steelhead trout (*Salmo gairdneri*).' *Can. J. Zool.*, *52*, 805-12.

Walker, B.W. (1949) 'Periodicity of spawning in the grunion, *Leuresthes tenuis*.' PhD thesis, University of California, Los Angeles, California.

Walker, M.G. (1971) 'Effect of starvation and exercise on the skeletal muscle fibres of the cod (*Gadus morhua* L.) and the coalfish (*Gadus virens* L.) respectively.' *J. Cons. Perm. Int. Explor. Mer.*, *33*, 421-6.

Walker, M.M., A.E. Dizon and J.L. Kirschvink (1982) 'Geomagnetic field detection by yellowfish tuna.' *IEEE Council on Oceanic Engineering*, pp. 755-8.

Walker, T.J. and A.D. Hasler (1949) 'Detection and discrimination of odors of aquatic plants by the bluntnose minnow (*Hyborhynchus notatus*, Raf.).' *Physiol. Zool.*, *22*, 45-63.

Waterman, T.H. and R.B. Forward (1970) 'Field evidence for polarized light sensitivity

in the fish *Zenarchopterus.' Nature*, *228*, 85-7.
Waterman, T.H. and R.B. Forward (1972) 'Field demonstration of polartaxis in the fish *Zenarchopterus*.' *J. Exp. Zool.*, *180*, 33-54.
Waterman, T.H. and W.E. Westell (1956) 'Quantitative effect of the sun's position on submarine light polarization.' *J. Mar. Res.*, *15*, 149-69.
Watson, T.A., B.A. McKeown, G.H. Geen, J.F. Powell and D.B. Parker (1984) 'The effect of pH on plasma electrolytes, carbonic anhydrase and ATPase activities in rainbow trout (*Salmo gairdneri*).' Water Res. (In review).
Weber, D. and G.J. Ridgway (1967) 'Marking Pacific salmon with tetracycline antibiotics.' *J. Fish. Res. Board Can.*, *24*, 849-65.
Wedemeyer, G. (1969) 'Stress-induced ascorbic acid depletion and cortisol production in two salmonid fishes.' *Comp. Biochem. Physiol.*, *29*, 1247-51.
Wedemeyer, G.A., R.L. Saunders and W.C. Clarke (1980) 'Environmental factors affecting smoltification and early sea water survival of anadromous salmonids.' *Mar. Fish. Rev.*, *6*, 1-14.
Weil, C., B. Breton and P. Reinaud (1975) 'Etude de la résponse hypophysaire à l'administration de Gn-Rh exogène au cours du cycle reproducteur annual chez la Carpe *Cyprinus carpio* L.' *CR Acad. Sci.*, *280*, 2469-72.
Wendelaar Bonga, S.E. and J.A.A. Greven (1975) 'A second cell type in Stannius bodies of two euryhaline teleost species.' *Cell. Tissue Res.*, *159*, 287-90.
White, H.C. (1934) 'A spawning migration of salmon in E. Apple River.' *Rep. Biol. Board Can.*, *19*, 33-41.
White, H.C. (1936) 'The homing of salmon in Apple River, NS.' *J. Fish. Res. Board Can.*, *2*, 391-400.
Whitehead, C., N.R. Bromage and J.R.M. Forster (1978) 'Seasonal changes in reproductive function of the rainbow trout (*Salmo gairdneri*).' *J. Fish. Biol.*, *12*, 601-8.
Wiebe, J.P. (1968) 'The affects of temperature and daylength on the reproductive physiology of the vivparous seaperch, *Cymatogaster aggregata* Gibbons.' *Can. J. Zool.*, *46*, 1207-19.
Wilhelmi, A.E. (1955) 'Comparative biochemistry of growth hormone from ox, sheep, pig, horse and fish pituitaries' in R.W. Smith Jr, O.H. Gaebler and C.N.H. Long (eds.), *The Hypophyseal Growth Hormone, Nature and Actions*, McGraw-Hill, New York, pp. 59-69.
Wilt, F.H. (1959) 'The organ specific action of thyroxine in visual pigmented differentiation.' *J. Embryol. Exp. Morph.*, *7*, 556-63.
Wingfield, J.C. and A.S. Grimm (1977) 'Seasonal changes in plasma cortisol, testosterone and oestradiol-17 β in the plaice *Pleuronectes platessa* L.' *Gen. Comp. Endocrinol.*, *31*, 1-11.
Winn, H.E., M. Salmon and N. Roberts (1964) 'Sun-compass orientation by parrot fishes.' *Z. Tierpsychol.*, *21*, 798-812.
Winters, G.H. (1970) 'Biological changes in coastal capelin from the over-wintering to the spawning condition.' *J. Fish. Res. Board Can.*, *27*, 2215-24.
Woo, N.Y.S., H.A. Bern and N.S. Nishioka (1978) 'Changes in body composition associated with smoltification and premature transfer to seawater in coho salmon (*Oncorhynchus kisutch*) and king salmon (*Oncorhynchus tschawytscha*).' *J. Fish. Biol.*, *13*, 420-8.
Woo, N.Y.S., W.C.M. Tong and E.L.P. Chan (1980) 'Effects of urophysical extracts on plasma electrolytes and metabolite levels in *Ophiocephalus maculatus*.' *Gen. Comp. Endocrinol.*, *41*, 458-66.
Wood, C.M. and D.J. Randall (1973) 'The influence of swimming activity on sodium balance in the rainbow trout (*Salmo gairdneri*).' *J. Comp. Physiol.*, *82*, 207.
Wood, J.D., D.W. Duncan and M. Jackson (1960) 'Biochemical studies on sockeye salmon during spawning migration. XI. The free histidine content of the tissues.' *J. Fish. Res. Board Can.*, *17*, 347-51.
Woodhead, A.D. (1975) 'Endocrine physiology of fish migration.' *Oceanogr. Mar. Biol. Ann. Rev.*, *13*, 287-382.
Woodhead, A.D. and P.M.J. Woodhead (1965) 'Seasonal changes in the physiology of the Barents Sea cod, *Gadus morhua* L., in relation to its environment.' *ICNAF Spec. Publ.*, *6*, 691-715.

Woodhead, P.M.J. (1970) 'An effect of thyroxine upon the swimming of cod.' *J. Fish. Res. Board Can.*, *27*, 2337-8.
Woodhead, P.M.J. (1966) 'The behavior of fish in relation to light in the sea.' *Oceanogr. Mar. Biol. Ann. Rev.*, *4*, 337-403.
Yamamoto, T. (1969) 'Sex differentiation' in W.S. Hoar and D.J. Randall (eds.), *Fish Physiology*, *Vol. III*, Academic Press, New York.
Yamazaki, F. (1976) 'Application for hormones in fish culture.' *J. Fish. Res. Board. Can.*, *33*, 948-58.
Yaron, Z., M. Cocos and H. Salzer (1980) 'Effects of temperature and photperiod on ovarian recrudescence in the cyprinid fish *Mirogrex terrae-sanctae*.' *J. Fish. Biol.*, *16*, 371-82.
Zambrano, D., R.S. Nishioka and H.A. Bern (1972) *The Innervation of the Pituitary Gland of Teleost Fishes*, K.M. Knigge, D.E. Scott and A. Windtl (eds.), Skarger, Basel.
Zaugg, W.S. and L.R. McLain (1976) 'Influence of water temperature on gill Na^+-K^+-ATPase activity in juvenile coho salmon (*Oncorhynchus kisutch*).' *Comp. Biochem. Physiol.*, *54A*, 419-21.
Zelnick, P.R. and K. Lederis (1973) 'Chromatographic separation of urotensins.' *Gen. Comp. Endocrinol.*, *20*, 392-400.
Zimmerman, M.A. and J.D. McCleave (1975) 'Orientation of elvers of American eels (*Anguilla rostrata*) in weak magnetic and electric fields.' *Helgoländer wiss. Meeresunters*, *27*, 175-89.

SUBJECT INDEX

SPECIES INDEX